The Pocket Atlas of Anatomy and Physiology

A Concise Reference for Students

Ruth Hull

T0075556

HUMAN
KINETICS

First published in 2024 by
Lotus Publishing
Apple Tree Cottage, Inlands Road, Nutbourne, Chichester, PO18 8RJ, and
Human Kinetics
1607 N. Market Street, Champaign, Illinois 61820

United States and International
Website: **US.HumanKinetics.com**
Email: info@hkusa.com
Phone: 1-800-747-4457

Canada
Website: **Canada.HumanKinetics.com**
Email: info@hkcanada.com

Illustrations Amanda Williams, Vicky Slegg
Text Design Medlar Publishing Solutions Pvt Ltd., India
Cover Design Chris Fulcher
Printed and Bound Kultur Sanat Printing House, Turkey

Disclaimer
This publication is written and published to provide accurate and authoritative information relevant to the subject matter presented. It is published and sold with the understanding that the author and publisher are not engaged in rendering legal, medical, or other professional services by reason of their authorship or publication of this work. If medical or other expert assistance is required, the services of a competent professional person should be sought.

British Library Cataloging-in-Publication Data
A CIP record for this book is available from the British Library

Library of Congress Cataloging-in-Publication Data
Names: Hull, Ruth, author.
Title: The pocket atlas of anatomy and physiology / Ruth Hull.
Description: First edition. | Chichester, England : Lotus Publishing ;
    Champaign, Illinois : Human Kinetics, 2024. | Includes index.
Identifiers: LCCN 2023016065 (print) | LCCN 2023016066 (ebook) |
    ISBN 9781718227040 (paperback) | ISBN 9781718227057 (epub) |
    ISBN 9781718227064 (pdf)
Subjects: MESH: Anatomy | Physiological Phenomena | Atlas | Handbook
Classification: LCC QP34.5 (print) | LCC QP34.5 (ebook) | NLM QS 17 |
    DDC 612--dc23/eng/20230510
LC record available at https://lccn.loc.gov/2023016065
LC ebook record available at https://lccn.loc.gov/2023016066

ISBN: 978-1-7182-2704-0
10 9 8 7 6 5 4 3 2 1 0

Contents

Introduction 6
*A Note on American
and British Spellings.* 8

**Chapter 1—Before You
Begin** 9
Anatomical Position 9
  Directional Terms. 9
Quadrants **11**
Anatomical Regions. **12**
Body Cavities **14**
Planes **16**

**Chapter 2—Organization
of the Body** **17**
**Levels of Structural
  Organization of the Body.** . . . **17**
**Chemical Organization
  of the Body** **18**
**Cellular Organization
  of the Body** **26**
  Transport across the Plasma
    Membrane 29
Life Cycle of a Cell **32**
**Tissue Level of Organization
  of the Body** **34**
**System Level of Organization
  of the Body** **43**
  Theory in Practice 46

**Chapter 3—The Skin, Hair,
and Nails** **47**
Skin **47**
Types of Skin **50**
Hair **52**
  Life Cycle of a Hair. 53
Nails. **54**
Cutaneous Glands. **55**
  Theory in Practice 56

**Chapter 4—The Skeletal
System** **57**
Bones **57**
  Bone Tissue 57
  Compact (Dense) Bone Tissue. . 58
  Spongy (Cancellous)
    Bone Tissue 59
  Bone Formation and
    Remodeling 59
Types of Bones. **60**
  Structure of a Long Bone. . . . 62
Organization of the Skeleton . . . **64**
Axial Skeleton **66**
  Bones of the Skull. 66
  Bones of the Neck and Spine . . 68
  Bones of the Thorax 72
Appendicular Skeleton. **74**
  Bones of the Upper Limb
    and Shoulder Girdle 74
  Bones of the Pelvic Girdle . . . 76
  Bones of the Lower Limb 77
  The Arches of the Foot 78
Joints **80**
  Synovial Joints 82
  Theory in Practice 88

**Chapter 5—The Muscular
System** **89**
Muscle Tissue **89**
Skeletal Muscle **91**
  Muscle Contraction
    and Relaxation. 94
  Types of Muscular
    Contraction 96
  Skeletal Muscles and
    Movement 97
Muscles of the Body **99**
  Muscles of the Face and Scalp. .104
  Muscles of the Neck 108
  Muscles of the Trunk
    and Shoulder 112

Muscles of the Back 117
Muscles of the Arm and
    Forearm 121
Muscles of the Hand 126
Muscles of the Hip and
    Thigh 128
Muscles of the Leg 132
Muscles of the Foot. 134
Principal Skeletal Muscles . . .137
Theory in Practice 138

**Chapter 6—The Nervous
System** **139**
**Organization of the Nervous
    System.** **140**
Nervous Tissue **143**
Structure of a Motor Neuron. . . .**145**
**Transmission of a Nerve
Impulse** **147**
Nerves. **148**
Brain. **148**
    Protection of the Brain 156
Cranial Nerves. **158**
Spinal Cord **160**
    Spinal Nerves 162
Special Sense Organs. **164**
    The Eye 164
    The Ear. 167
    The Mouth. 170
    The Nose. 171
    Theory in Practice 172

**Chapter 7—The Endocrine
System** **173**
**Endocrine Glands and
    Their Hormones** **174**
    Hormones Controlled
        by the Pituitary Gland 183
    Hormones not Controlled
        by the Pituitary Gland 184
The Stress Response **185**
    Theory in Practice 188

**Chapter 8—The Respiratory
System** **189**
**Organization of the Respiratory
    System.** **190**
    The Nose. 192
    Paranasal Sinuses. 193
    Pharynx (Throat) 193
    Larynx (Voice Box) 194
    Trachea (Windpipe) 195
    Bronchi. 196
    Lungs 197
    Alveoli (Air Sacs) 198
Physiology of Respiration **198**
    Pulmonary Ventilation 198
    External Respiration
        (Pulmonary Respiration). . .199
    Internal Respiration (Tissue
        Respiration) 199
    Theory in Practice 202

**Chapter 9—The Cardiovascular
System** **203**
Blood **203**
Heart **207**
    Walls, Chambers, and
        Valves of the Heart 209
    Blood Flow to the
        Heart Tissue 211
    Blood Flow Through
        the Heart 213
    Physiology of the Heart 214
Blood Vessels **215**
Blood Pressure. **217**
**Primary Blood Vessels
    of Systemic Circulation** **218**
    Primary Arteries of the
        Head, Face, and Neck. . . .220
    Primary Veins of the
        Head, Face, and Neck. . . .221
    Guide to Flowcharts of the
        Blood Vessels of the Body . .221
    Primary Arteries of the
        Upper Limbs. 222

Primary Veins of the
  Upper Limbs. 223
Primary Arteries of the
  Thorax 224
Primary Veins of the
  Thorax 225
Primary Arteries of the
  Abdomen. 226
Primary Veins of the
  Abdomen. 227
Primary arteries of the Pelvis
  and Lower Limbs 230
Primary Veins of the Pelvis
  and Lower Limbs 231
Theory in Practice 232

**Chapter 10—The Lymphatic
and Immune System 233**
**Organization of the Lymphatic
and Immune System. 235**
  Lymph Nodes. 238
**Resistance to Disease
and Immunity 241**
  Non-Specific Resistance
   to Disease 241
  Immunity (the Immune
   Response) 242
  Theory in Practice 244

**Chapter 11—The Digestive
System 245**
**Organization of the Digestive
System. 247**
  Peritoneum 247
  Walls of the GI Tract 248
**Gastrointestinal Tract and Its
Accessory Organs 250**

Mouth (Oral or Buccal
  Cavity) 250
Esophagus. 253
Stomach 253
Liver 256
Gall Bladder. 258
Small Intestine 258
Pancreas 263
Large Intestine 264
Digestion of Carbohydrates . . .266
Digestion of Proteins 267
Digestion of Lipids 267
Theory in Practice 268

**Chapter 12—The Urinary
System 269**
  Kidneys 270
  Nephron 272
  Ureters. 274
  Urinary Bladder 274
  Urethra. 275
  Theory in Practice 276

**Chapter 13—The Reproductive
System 277**
Male Reproductive System 278
Female Reproductive System . . .281
  Mammary Glands 284
  Female Reproductive Cycle . . .286
**Aging and the Reproductive
  System. 290**
  Theory in Practice 292

Glossary 293
Index 319

Introduction

This book is for anyone studying to be a healthcare professional. It presents all the information necessary to gain a thorough understanding of the subject in a clear, accurate and easily absorbed format. We have tried to strike a balance between a friendly, informal tone and serious academic content.

We hope you will enjoy using this book and would welcome any feedback, good or bad, which will help us to improve it in subsequent editions.

General Editor

Maia Vaswani studied microbiology and genetics and human osteoarchaeology, and then took up a position as research assistant in the Entomology Department of the Natural History Museum in London, where she worked for several years.

After brief sojourns in Budapest and Canada, she settled in Brittany and took up freelance copy-editing about ten years ago.

She has recently moved to a cottage in Wales, where she lives with her partner and three cats.

Additional Material

The publisher would like to thank Dr. Daniel Quemby (MBBS (Hons), BSc (Hons), BSc. Med. Sci, MRCS, FRCA, medical doctor, anesthetist and expedition doctor) for his kind permission to use some additional material that has been included in Chapters 4 and 5. This material first appeared in *The Concise Book of Muscles, Fourth Edition*, also published by Lotus.

Abbreviations

AIIS	Anterior inferior iliac spine
ANS	Autonomic nervous system
ASIS	Anterior superior iliac spine
CNS	Central nervous system
IP	Interphalangeal
ITT or ITB	Iliotibial tract or iliotibial band
MCP	Metacarpophalangeal
MTP	Metatarsophalangeal
PNS	Peripheral nervous system

A Note on American and British Spellings

This book uses US spelling conventions, and some of the technical ones may be unfamiliar to students from the UK or elsewhere. Students should be aware of these different spellings as they may encounter them in their future studies or work.

Examples include the simple s/z difference, such as in "mineralised" and "catalysed" (UK) versus "mineralized" and "catalyzed" (US). For the most part, the differences in spelling are minor; for example, "ae" in the British "anaemia" and "haemoglobin" versus just "e" in the US "anemia" and "hemoglobin." Or "oe" in the British "oestrogen" and "oesophagus" versus just "e" in the American "estrogen" and "esophagus."

There is an exception, however, where the most common term in the US is an entirely different word from the one most commonly used in the UK: "epinephrine" is the preferred American form for what the British mostly call "adrenaline." The same difference is seen with "norepinephrine" and "noradrenaline." This book uses "epinephrine" and "norepinephrine," but in places readers will see "adrenaline" and "noradrenaline" also, as a reminder.

CHAPTER 1

Before You Begin

Anatomical Position

The anatomical position is a basic position that can always be used as a reference point (figure 1.1).

- Head facing forward
- Feet parallel
- Arms hanging by the side
- Most importantly, palms facing forward.

The front of the body (where the face is) is called the **anterior** or **ventral**, and the back of his body is the **posterior** or **dorsal**.

Directional terms
The **midline** or **median line** is an imaginary line that acts as a reference for many anatomical terms, including those describing certain movements.

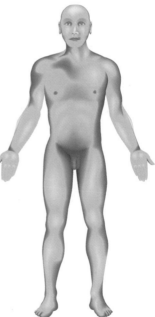

Figure 1.1: Anatomical position

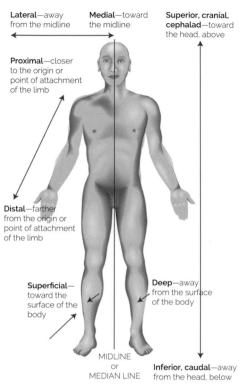

Figure 1.2: Directional terms

In relation to the midline are these additional directional terms:

Abduction movement away from the midline

Adduction movement toward the midline

Anterior toward the front of the body, in front of

Dorsal at the back of the body, behind

Extension movement at a joint increasing the angle between body parts

Flexion movement at a joint decreasing the angle between body parts

Palmar anterior surface of the hand

Plantar sole of the foot

Posterior at the back of the body. behind

Ventral at the front of the body, in front of

Quadrants

The abdominopelvic cavity is large and contains many organs. It is, therefore, helpful to divide it into smaller regions using imaginary lines that can be named according to their relative positions. These regions are **quadrants** and include the right upper quadrant (RUQ), the left upper quadrant (LUQ), the right lower quadrant (RLQ) and the left lower quadrant (LLQ).

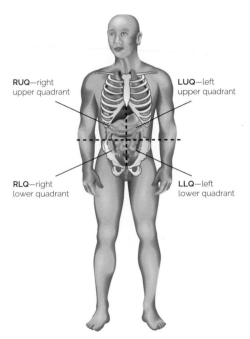

RUQ—right upper quadrant

LUQ—left upper quadrant

RLQ—right lower quadrant

LLQ—left lower quadrant

Figure 1.3: Quadrants

Anatomical Regions

These relate to specific areas of the body. For example, the neck is the cervical region. When we discuss cervical nerves or cervical vertebrae we are talking about the nerves and vertebrae of the neck.

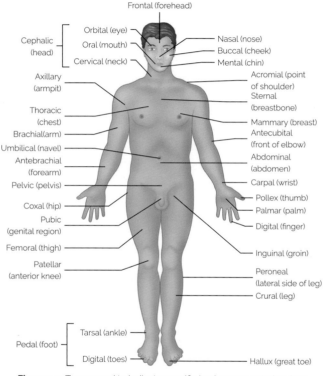

Figure 1.4: Terms used to indicate specific body areas, anterior view

The two primary divisions of the body are its *axial* parts, consisting of the head, neck, and trunk, and its *appendicular* parts, consisting of the limbs, which are attached to the axis of the body. Figures 1.4 and 1.5 show the terms used to indicate specific body areas. Terms in parentheses are the lay terms for the area.

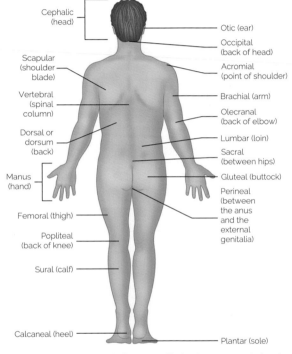

Figure 1.5: Terms used to indicate specific body areas, posterior view

Body Cavities

Body cavities are spaces within the body that contain and protect the internal organs. For example, the cranial cavity contains the brain. There are two main cavities:

1. The **dorsal** cavity (at the back of the body)
2. The **ventral** cavity (at the front of the body).

These are subdivided as follows:

Dorsal Cavity	
Cranial cavity	Contains the brain and is protected by the bony skull (cranium)
Spinal cavity/canal	Contains the spinal cord and is protected by the vertebrae

Ventral Cavity	
Thoracic cavity	Contains the trachea, two bronchi, two lungs, the heart and the esophagus, and is protected by the ribcage; it is separated from the abdominal cavity by the diaphragm
Abdominopelvic cavity, consisting of the:	
Abdominal cavity	Contains the stomach, spleen, liver, gall bladder, pancreas, small intestine and most of the large intestine, covered by a serous membrane called the peritoneum; the abdominal cavity is mainly protected by the muscles of the abdominal wall and partially by the ribcage and diaphragm
Pelvic cavity	Contains a portion of the large intestine, the urinary bladder and the reproductive organs; it is protected by the pelvic bones

Note: For ease of learning, blood vessels, lymphatic vessels, lymph nodes and nerves have not been included in these charts. When learning about the body cavities it is important to know the following terms:

*   **Parietal:** Relating to the inner walls of a body cavity
*   **Visceral:** Relating to the internal organs of the body.

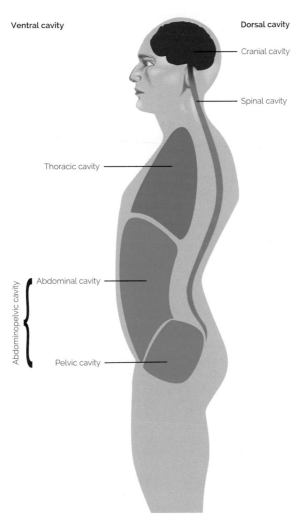

Figure 1.6: Body cavities

Planes

Anatomists often divide the body or an organ into sections so that they can study its internal structures. These sections are made along imaginary flat surfaces called **planes**, which are as follows:

- **Sagittal plane:** Divides the body vertically into right and left portions
- **Frontal/coronal plane:** Divides the body vertically (longitudinally) into posterior and anterior portions
- **Transverse plane (cross-section):** Divides the body horizontally into inferior and superior portions
- **Oblique plane:** Divides the body at an angle between the transverse plane and the frontal/sagittal plane.

Sagittal plane

Frontal/coronal plane

Transverse plane

Oblique plane

Figure 1.7: Planes

CHAPTER 2

Organization of the Body

Levels of Structural Organization of the Body

Chemical level: The body is made up of tiny building blocks called atoms. Atoms such as carbon, hydrogen and oxygen are essential for maintaining life and combine to form molecules such as fats, proteins and carbohydrates. These molecules then combine to form cells.

Cellular level: Cells are the basic structural and functional unit of the body. There are many different types of cells and they all have very specific functions. For example, there are nerve cells and blood cells. Cells combine to form tissues.

Tissue level: Tissues are groups of cells and the materials surrounding them. There are four tissue types, which perform particular functions. These types are epithelial, muscular, connective and nervous tissue. Two or more different types of tissue combine to form organs.

Organismic level: An organism is a living person – you.

Organ level: Organs have very specific functions and recognizable shapes. Examples of organs include the heart, liver, stomach and lungs. Organs that share a common function combine to form systems.

System level: Systems are composed of related organs and they perform a particular function. For example, the digestive system is composed of organs such as the stomach, liver, pancreas, and small and large intestines. Its function is to break down and digest food. Finally, systems combine to form the organism.

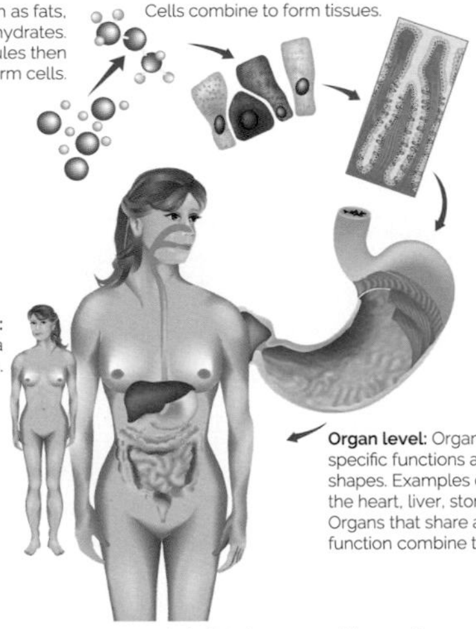

Figure 2.1: Levels of structural organization of the body

Chemical Organization of the Body

Table 2.1: Glossary of basic chemistry terms

Matter	Anything that occupies space (has volume) and has mass (weight). All living and non-living things consist of matter
Atom	The smallest unit of matter that retains the properties of that matter. An atom can be broken down into subatomic particles called protons, neutrons, and electrons, which are themselves made of smaller particles
Element	A substance made up of only one type of atom. The number of protons in the nucleus of the atom defines the element, and never changes
Ion	A "charged" atom. Although the number of protons (*atomic number*) of an atom can never change, the number of electrons can change. An atom can lose or gain another electron, and this will change the charge of the atom
Anion	A negative ion. The atom has more electrons than protons and hence a net negative charge
Cation	A positive ion. The atom has lost one or more of its electrons, giving it a net positive charge
Molecule	The combination of two or more atoms
Compound	A molecule containing atoms of two or more different elements
Chemical reaction	Occurs when atoms combine with or break apart from other atoms to form new products
Chemical bond	The force of attraction that holds atoms or ions together. The number of protons in the nucleus of the atom defines the element, and never changes. The electrons, however, can be attracted to other atoms and interact (bond) with them. When atoms bind with one another, it is the distribution of their valence electrons that determines the chemical bonds that are formed. Valence electrons are the electrons found in the outermost electron shell of the atom There are three types of chemical bonds:

Table 2.1: (continued)

	Ionic bonds	The electrons from one atom are fully transferred to the outer shell of another atom. This leaves the atoms with an "imbalanced" number of electrons and protons and so the atoms are now charged—they are now ions and are held together through an electrostatic attraction
	Covalent bonds	Two atoms share pairs of electrons and form molecules
	Hydrogen bonds	Form between hydrogen atoms and other atoms because of the attraction of oppositely charged molecules, rather than the actual sharing or transfer of electrons. They are weak bonds that do not bind atoms into molecules. Hydrogen bonds link water molecules together
pH	A measure of the concentration of hydrogen ions (H^+) in a solution: the higher the concentration of hydrogen ions, the lower the pH and the more acidic it is. Alkalis contain hydroxide ions (OH^-), which in solution will react with free hydrogen ions to form water, thereby lowering the concentration of hydrogen ions, raising the pH, and making the solution alkaline. pH is measured on a scale from 0 to 14, with zero being very acidic (contains more H^+), 14 being alkaline (contains more OH^-), and 7 being neutral	
Salt	A substance that when dissolved in water dissociates into cations and anions, none of which are hydrogen ions or hydroxide ions	
Electrolyte	A charged particle (ion) that conducts an electrical current in an aqueous solution	

The periodic table displays chemical elements arranged from left to right and top to bottom in order of increasing atomic number (number of protons).

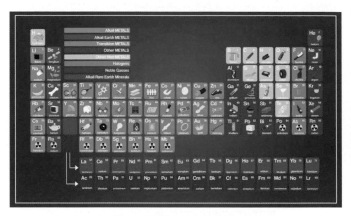

Figure 2.2: The periodic table

The pH scale measures the relative acidity or alkalinity of a solution on a scale from 0 to 14, with zero being acidic, 7 neutral, and 14 alkaline.

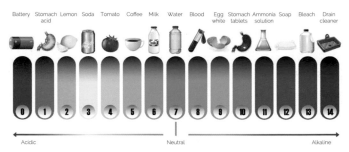

Figure 2.3: The pH scale

Table 2.2: Elements of the body

Element	Role in the Body
There are four major elements in the body, which make up 96% of the body's mass. They are:	
Oxygen	A component of water and organic molecules and essential to cellular respiration, a process in which cellular energy, adenosine triphosphate (ATP), is produced
Carbon	The main component of all organic molecules (e.g., carbohydrates, lipids, proteins, and nucleic acids)
Hydrogen	A component of water, all foods, and organic molecules; also influences the pH of body fluids
Nitrogen	A component of all proteins and nucleic acids
There are nine lesser elements in the body, which make up 3.9% of the body's mass. They are:	
Calcium	Found in bones and teeth; also necessary for muscle contraction, nerve transmission, release of hormones, and blood clotting
Phosphorus	Found in bones and teeth as well as in nucleic acids and many proteins; also forms part of ATP
Potassium	Necessary for many chemical reactions within the cell; also important for nerve impulses and muscle contraction
Sulfur	A component of some vitamins and many proteins
Sodium	Necessary for many chemical reactions in the extracellular fluid (fluid outside of the cell); also plays a role in water balance, nerve impulses, and muscle contraction
Chlorine	Necessary for many chemical reactions in the extracellular fluid
Magnesium	Found mainly in bone and necessary for the activity of more than 300 enzymes in the body
Iodine	Necessary for the synthesis of thyroid hormones
Iron	A component of the hemoglobin molecule, which transports oxygen within red blood cells
There are thirteen other elements in the body, which are present in such small quantities that they are known as trace elements. They make up 0.1% of the body's mass and are:	
Aluminum, boron, chromium, cobalt, copper, fluorine, manganese, molybdenum, selenium, silicon, tin, vanadium, and **zinc**	

Table 2.3: Major compounds of the body

Inorganic compounds: *These compounds generally do not contain the element carbon and are usually simpler and smaller than organic compounds*		
Compound	**Elements Present**	**Role in the Body**
Water	Hydrogen, oxygen	This is the most abundant substance in the body because it is the *solvent* in body fluids, meaning that different materials and substances can dissolve in it
Water has many important functions in the body, including:		
Maintaining body temperature	Water can absorb and give off large amounts of heat without its temperature changing too much; it is therefore able to maintain a normal internal temperature despite hot sun, cold winds and other external changes	
Acting as a lubricant	Water acts as a lubricant where internal organs touch and slide over one another or where bones, ligaments, and tendons meet and rub together; it also lubricates the gastrointestinal tract so that food can easily move through it and feces out of it	
Providing cushioning	Water creates a protective cushion around certain organs; e.g., it forms cerebrospinal fluid, which cushions the brain and protects it from external trauma	
Being a "universal solvent"	Water can dissolve or suspend many different substances and is therefore an ideal medium in which chemical reactions can take place; it can also transport substances such as nutrients, respiratory gases, and waste around the body	
Organic compounds: *These are compounds that* **contain the element carbon.** *Carbon is a very useful element because it reacts easily to form large molecules that do not dissolve easily in water. These molecules build body structures and also break down to give off energy when the body needs it. The compounds below are all organic compounds and are important nutrient chemicals to the body*		

Table 2.3: *(continued)*

Compound	Elements Present	Role in the Body
Carbohydrates	Carbon, hydrogen, oxygen	Carbohydrates are the fuel of the body
Carbohydrates include:		
Monosaccharides and disaccharides (simple sugars)	Simple sugars are the building blocks of carbohydrates and a source of energy for chemical reactions Glucose is the main form in which sugar is used by your cells, fructose is found in fruits, and galactose is present in milk	
Some sugars also form parts of structural units, e.g., deoxyribose is a sugar that forms part of the DNA molecule, which carries hereditary information		
Starch	This is a large molecule and is the main carbohydrate found in food	
Glycogen	Glycogen is an energy reserve and is stored in the liver and skeletal muscles	
Cellulose	This is a carbohydrate built by plants that we eat but cannot digest, and so it creates bulk which aids the movement of food and waste through our intestines; it is commonly referred to as fiber	
Lipids (fats)	Carbon, hydrogen, oxygen	Lipids have over double the energy value of carbohydrates and proteins and are easily converted to body fat; they are composed of glycerol and fatty acids and most of them are *hydrophobic* (insoluble in water)
There are many different types of lipids with very diverse roles, the main ones being:		
Triglycerides (neutral fats)	These are the most plentiful fats in the body and in your diet and are solids (fats) or liquids (oils) at room temperature; they are found in fat deposits beneath the skin and around organs; they protect and insulate the organs and are also a major source of stored energy	

Table 2.3: (continued)

Compound	Elements Present	Role in the Body
Phospholipids	These lipids form an integral part of the cell membrane and are also found in high concentrations in the nervous system	
Steroids	Steroids are a type of lipid and many different types of steroids are found in the body; e.g., sex hormones and cholesterol (a steroid-alcohol or sterol)	

Other types of lipids include **eicosanoids**, which have diverse effects on inflammation, immunity, blood clotting, and other bodily responses; **fatty acids**, which are important energy-supplying molecules; **carotenes**, needed for the synthesis of vitamin A; **vitamin E**, which contributes to the functioning of the nervous system, wound healing, and is also an antioxidant; **vitamin K**, which is necessary for blood clotting; and **lipoproteins**, which help transport lipids in the body

Compound	Elements Present	Role in the Body
Proteins	Carbon, hydrogen, oxygen, nitrogen; may contain sulfur	Proteins are the main family of molecules from which the body is built and are themselves built from amino acids; they are diverse in size, shape, and function

Proteins play the following roles in the body:

Structural	Proteins are the building blocks of the body; e.g., bone is built from collagen, and skin, hair, and nails are built from keratin
Regulatory	Hormones are made from proteins, and they regulate the bodily functions; e.g., insulin helps regulate blood glucose levels
Contractile	The proteins myosin and actin enable muscles to shorten, which allows for movement
Immunity	Antibodies are proteins that protect against invading microbes
Transport	Some carrier molecules are proteins; e.g., hemoglobin transports oxygen in the blood
Catalytic	**Enzymes** are proteins that help speed up biochemical reactions
Energy source	Proteins can also act as a source of energy in times of dietary inadequacy

Table 2.3: (continued)

Compound	Elements Present	Role in the Body
Proteins are the main family of molecules from which the body is built and are themselves built from amino acids; the standard genetic code encodes for 20 different amino acids, 9 of which are called essential amino acids because they cannot be made by the body and therefore must be obtained from proteins in the diet		
Nucleic acids	Carbon, hydrogen, oxygen, nitrogen, phosphorus	Nucleic acids are very important molecules that are found inside cells
There are two types of nucleic acids:		
Deoxyribonucleic acid (DNA)	DNA is found inside the cell's nucleus and makes up chromosomes, which contain our genes	
	DNA is the inherited genetic material inside every cell; it provides instructions for building every protein in the body and it replicates itself before a cell develops to ensure that the genetic information inside every body cell is identical	
Ribonucleic acid (RNA)	RNA is the "molecular slave" to DNA (Marieb, 2003, p. 47); it transports the orders of DNA from the nucleus to ribosomes, where it is used to create specific proteins as per the genetic code	
Adenosine triphosphate (ATP)	Carbon, hydrogen, oxygen, nitrogen, phosphorus	ATP is the main energy-transferring molecule in the body and provides a form of chemical energy that can be used by all body cells

Cellular Organization of the Body

Cytology is the study of cells, which are the basic structural and functional units of the body.

Cells vary greatly in size, shape, and structure according to their function, but despite their differences, cells are bathed in **interstitial fluid**, a dilute saline solution derived from the blood. Interstitial fluid is outside of the cell and is also known as extracellular fluid, intercellular fluid, or tissue fluid. The fluid inside the cells is called **intracellular fluid**.

Both the interstitial fluid and the intracellular fluid are made up of oxygen, nutrients, waste, and other particles dissolved in water.

Table 2.4: The structure of a generalized animal cell

Cell Structure	Function	
Plasma membrane	A thin and flexible barrier that surrounds the cell and regulates the movement of all substances into and out of it. It is made up of lipids and proteins	
	Three types of lipids are present in the plasma membrane:	
	Phospholipids	Each phospholipid molecule has a hydrophilic (water-loving) head and a hydrophobic (water-hating) tail. The phospholipids lie tail to tail in two parallel layers to form the phospholipid bilayer, which is the framework of the membrane
	Glycolipids	Play a part in cell communication, growth, and development
	Cholesterol	Is found only in animal cells and helps to strengthen the membrane
	Two types of proteins are scattered in the phospholipid bilayer:	
	Integral proteins	Extend all the way through the membrane to create channels, which allow for the passage of materials in and out of the cell
	Peripheral proteins	Are loosely attached to the surfaces of the membrane and can separate easily from it

Table 2.4: (continued)

Cell Structure	Function
Cytoplasm	All the cellular material inside the plasma membrane, excluding the nucleus. It is the site of most cellular activities and includes cytosol, organelles, and inclusions
Cytosol	Cytosol is a thick, transparent, gel-like fluid made up of mainly water. It also contains solids and solutes, as well as spaces called vacuoles, which house cellular wastes and secretions
Organelles are the "little organs" of the cell and they play a specific role in maintaining the life of the cell. They are very specialized structures and differ according to the cell type and its functions. Some of the important organelles are:	
Nucleus	Controls all cellular structure and activities. It is surrounded by a nuclear envelope and contains the nucleolus where ribosomes are assembled; the chromatin, which is a mass of chromosomes tangled together in non-dividing cells; and genes, which are the hereditary units of the cell and control its structure and activity. Genes are arranged along structures called chromosomes
Mitochondria	The "powerhouses" of the cell where ATP is generated through the process of cellular respiration
Ribosomes	The sites of protein synthesis in the cell. Some ribosomes float freely in the cytoplasm and are called free ribosomes. Others are attached to the endoplasmic reticulum
Endoplasmic reticulum (ER)	Provides a surface area for chemical reactions and transports molecules within the cell. There are two types of endoplasmic reticulum: rough ER, which has ribosomes attached to it and provides a site for protein synthesis, temporarily stores new protein molecules, participates in the formation of glycoproteins, and works together with the Golgi complex to make and package molecules; and smooth ER, which has no ribosomes attached to it and provides a site for the synthesis of certain lipids and the detoxification of various chemicals
Golgi complex/ apparatus	Processes, sorts, and packages proteins and lipids for delivery to the plasma membrane. It also forms lysosomes and secretory vesicles

Table 2.4: (continued)

Cell Structure	Function
Lysosomes	Vesicles formed inside the Golgi complex that contain digestive enzymes and function in recycling different molecules
Peroxisomes	Vesicles containing enzymes that detoxify any potentially harmful substances in the cell
Centrosomes	In non-dividing cells they organize the microtubules that help support and shape the cell and move substances. In dividing cells they form the mitotic spindle. Centrosomes contain centrioles
Centrioles	Found within the centrosomes and play a role in cell division and also in the formation and regeneration of flagella and cilia

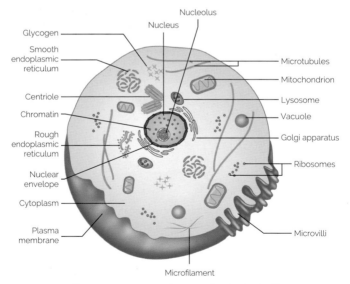

Figure 2.4: Structure of a generalized animal cell

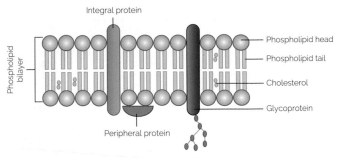

Figure 2.5: Structure of the plasma membrane

Transport across the plasma membrane

Materials are transported into and out of the cell via the plasma membrane, and there are two types of transport: **passive** processes and **active** processes. In passive processes substances are moved without using cellular energy. In active processes cells need to use some of their own energy gained from the splitting of ATP to transport materials across the plasma membrane. Active processes usually involve moving substances against a concentration gradient.

Passive processes

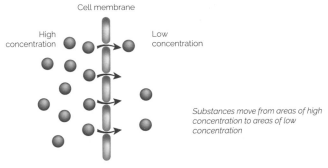

Substances move from areas of high concentration to areas of low concentration

Figure 2.6: Simple diffusion

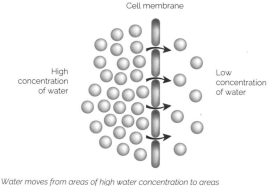

Cell membrane

High concentration of water

Low concentration of water

Water moves from areas of high water concentration to areas of lower water concentration, or ... water moves from areas of low solute concentration to areas of high solute concentration. Both of these processes are always across a selectively permeable membrane

Figure 2.7: Osmosis

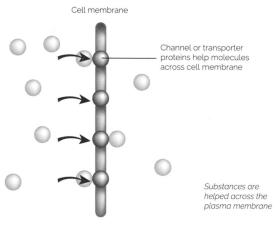

Cell membrane

Channel or transporter proteins help molecules across cell membrane

Substances are helped across the plasma membrane

Figure 2.8: Facilitated diffusion

Active processes

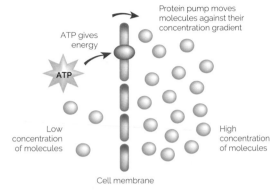

Protein pump moves molecules against their concentration gradient

ATP gives energy

ATP

Low concentration of molecules

High concentration of molecules

Cell membrane

Pumps powered by ATP push molecules across the plasma membrane

Figure 2.9: Active transport

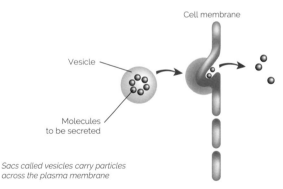

Cell membrane

Vesicle

Molecules to be secreted

Sacs called vesicles carry particles across the plasma membrane

Figure 2.10: Vesicular transport

Life Cycle of a Cell

A cell has two distinct periods in its life cycle: **interphase** and **cell division**. Interphase is the time in which a cell grows and carries out its usual metabolic activities. It is also the time in which the cell prepares itself for division through DNA replication. When DNA replicates, the molecule uncoils and separates into its two nucleotide strands. Each strand now serves as a template for building another nucleotide strand. This results in two identical double-stranded helices.

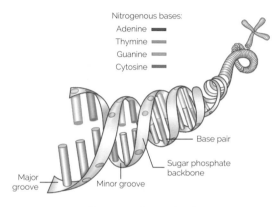

Nitrogenous bases:
Adenine
Thymine
Guanine
Cytosine

Base pair

Sugar phosphate backbone

Major groove

Minor groove

Figure 2.11: DNA molecule

Table 2.5: Types of cell division

Somatic Cell Division: Mitosis	Reproductive Cell Division: Meiosis
Cellular reproduction in which a mother cell divides into two daughter cells, each containing the same genes as the mother cell	Reproductive cell division in which four haploid daughter cells are produced
Key differences to remember:	
• Occurs in all cells, except bacteria and some cells of the reproductive system	• Only occurs in reproductive cells
• Takes place when the body needs to replace dead and injured cells or produce new cells for growth	• Takes place when a new organism is to be produced

Table 2.5: (continued)

Somatic Cell Division: Mitosis	Reproductive Cell Division: Meiosis
• Two identical diploid daughter cells are produced. Diploid cells have two complete sets of chromosomes per cell	• Four haploid daughter cells are produced. Haploid cells have only one set of chromosomes per cell. Human cells have 46 chromosomes. You inherit 23 chromosomes from your mother and 23 from your father

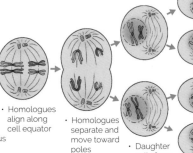

Mitosis

- Centrioles move toward poles
- Chromatin begins to form into chromosomes
- Nuclear envelope disintegrates

- Chromosomes align along cell equator to form metaphase plate

- Sister chromatids separate and move toward poles

- Daughter cells form
- Nuclei are genetically identical to parent cell

Meiosis

- Synapsis and crossing-over occur
- Paired homologous chromosomes
- Nuclear envelope disintegrates

- Homologues align along cell equator

- Homologues separate and move toward poles

- Daughter cells form

- Daughter chromosomes separate to form gametes
- Nuclei are not genetically identical to parent cell

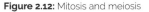

Figure 2.12: Mitosis and meiosis

Tissue Level of Organization of the Body

Similar cells working together to perform a common function form groups called tissues. For example, nerve cells work together to form nervous tissue. **Histology** is the study of tissues and there are four types in the body: epithelial, connective, muscle, and nervous.

Table 2.6: Types of tissue in the body

Name	Description	Function
Epithelial	Cells fit closely together with little extracellular material between them, are arranged in continuous sheets, and always have a free surface that is exposed to the body's exterior or a body cavity and an interior surface that is attached to a basement membrane. Cells are constantly being regenerated by mitosis and epithelial tissue (epithelium) is avascular and firmly joined to connective tissue	Covers body surfaces, lines hollow organs, cavities, and ducts, and forms glands. Functions in protection, absorption, filtration, and secretion. For example, the skin
Connective	Cells are surrounded and separated by a matrix made of protein fibers and a fluid, gel, or solid ground substance. This ground substance is usually produced by the connective tissue cells and deposited in the space between the cells. The tissue has a rich blood supply, except for cartilage and tendons, and has a nerve supply, except for cartilage	Protects and supports the body and its organs, binds organs together, stores energy reserves as fat, and provides immunity. For example, blood, adipose tissue (fat), and bone
Muscle	Fibers of muscle tissue are elongated cells that provide a long axis for contraction. Thus, muscle tissue is able to shorten (contract)	Provides movement and force, maintains posture, and generates heat. For example, skeletal muscle

Table 2.6: (continued)

Name	Description	Function
Nervous	Made up of neurons and neuroglia and found in the brain, spinal cord, and nerves. Neurons are composed of a cell body whose cytoplasm is drawn out into a long extension. These extensions are called dendrites or axons and they receive and conduct electrochemical impulses. Neuroglia are the supporting cells that insulate, support, and protect the neurons. They do not generate or conduct nerve impulses	Initiates and transmits nerve impulses. For example, nerves

Table 2.7: Classification of epithelial tissue

Simple epithelium: Simple epithelium is a single layer of cells. It is usually very thin and functions in absorption, secretion, and filtration. It includes:

	Simple squamous (pavement) epithelium	
	Description	A single layer of flat squamous cells
	Location example	Lines blood vessels, lymphatic vessels, and air sacs of lungs, and forms serous membranes
	Function	Filtration, diffusion, osmosis, and secretion
	Simple cuboidal epithelium	
	Description	A single layer of cube-shaped cells
	Location example	Covers the surface of the ovaries, lines kidney tubules, and forms the ducts of many glands
	Function	Secretion and absorption
	Simple columnar epithelium	
	Description	A single layer of rectangular cells; may contain goblet cells, which produce a lubricating mucus, and microvilli, which are finger-like projections that increase the surface area of the plasma membrane
	Location example	Lines the gastrointestinal tract
	Function	Secretion and absorption

Table 2.7: (continued)

	Ciliated simple columnar epithelium	
	Description	A single layer of rectangular cells that contain hairlike projections called cilia, which help move substances
	Location example	Lines part of the upper respiratory tract, the fallopian tubes, and some of the sinuses
	Function	Moves fluids or particles along a passageway

Stratified epithelium: Stratified epithelium consists of two or more layers of cells. It is durable and functions in protecting underlying tissues in areas of wear and tear. Some stratified epithelia also produce secretions. Stratified epithelium includes:

	Stratified squamous epithelium	
	Description	Consists of several layers of cells that are squamous in the superficial layer and cuboidal to columnar in the deep layers
		It exists in keratinized and non-keratinized forms; **keratin** is a tough, waterproof protein that is resistant to friction and helps repel bacteria
	Location example	The superficial layer of the skin is a keratinized form, while wet surfaces such as the mouth and tongue are non-keratinized
	Function	Protection
	Transitional epithelium	
	Description	Consists of layers of cells that change shape when the tissue is stretched
	Location example	Organs of the urinary system, e.g., the bladder
	Function	Allows for distension

Table 2.8: Classification of connective tissue

Loose connective tissue: *Loose connective tissue has loosely woven fibers and many cells. It is a softer connective tissue and includes:*			
	Areolar tissue		
		Description	This is a loose, soft, pliable, semifluid tissue and is the most widely distributed type of connective tissue in the body
			It contains collagen, elastic, and reticular fibers, which give areolar tissue its elasticity and extensibility (see chapter 3 for more information on these fibers)
		Location example	Surrounds body organs and is found in the subcutaneous layer of the skin
		Function	Strength, elasticity, and support (it is often the glue that holds the internal organs in their positions)
	Adipose tissue		
		Description	This is an areolar tissue that contains mainly fat cells
		Location example	Subcutaneous layer of the skin, surrounding organs such as the kidneys and heart, and found in the mammary glands
		Function	Insulation, energy reserve, support, and protection
	Lymphoid (reticular) tissue		
		Description	Contains thin fibers that form branching networks to form the **stroma** (a support framework for many soft organs)
		Location example	Lymph nodes, the spleen, red bone marrow
		Function	Support

Dense connective (fibrous) tissue: *Dense connective tissue contains thick, densely packed fibers and fewer cells than loose connective tissue. It forms strong structures, and there are different types of dense connective tissue, including the following:*

Table 2.8: (continued)

	Elastic connective tissue (yellow elastic tissue)	
	Description	Consists of many freely branching elastic fibers, few cells, and little matrix; the elastic fibers give the tissue a yellowish color
	Location example	Lung tissue, walls of arteries, trachea, bronchial tubes
	Function	Elasticity for stretching and strength
	Dense regular connective tissue (white fibrous tissue)	
	Description	Consists mainly of collagen fibers arranged in parallel bundles with cells in between the bundles; the matrix is a shiny white color
	Location example	Tendons, ligaments, aponeuroses
	Function	Attachment

Bone (osseous tissue): *Bone is an exceptionally hard connective tissue that protects and supports other organs of the body. It consists of bone cells sitting in **lacunae** (cavities) surrounded by layers of a very hard matrix that has been strengthened by inorganic salts such as calcium and phosphate. It includes compact and cancellous bone*

	Description	**Compact bone** is the hard, more dense, outer type of bone, while **cancellous bone** lies internally to the compact bone and is spongy
	Location example	Bones
	Function	Support, protection, movement, and storage

Cartilage: *Cartilage is a resilient, strong connective tissue that is less hard and more flexible than bone. It consists of a dense network of collagen and elastic fibers embedded in a rubbery ground substance. Cartilage has no blood vessels or nerves and includes:*

Table 2.8: *(continued)*

	Hyaline cartilage		
	Description	This is the most abundant cartilage in the body but is also the weakest It consists of a resilient, gel-like ground substance, fine collagen fibers, and cells	
	Location example	Ends of long bones, ends of ribs, nose	
	Function	Movement, flexibility, and support	
	Fibrocartilage		
	Description	This is the strongest of the three types of cartilage and consists of cells scattered between bundles of collagen fibers; a strong and rigid cartilage	
	Location example	Intervertebral disks	
	Function	Support and strength	
	Elastic cartilage		
	Description	This is a strong and elastic cartilage consisting of cells in a threadlike network of elastic fibers within the matrix	
	Location example	Eustachian tubes, supports external ear and epiglottis	
	Function	Supports and maintains shape	

Blood (vascular tissue): Blood is an unusual type of connective tissue whose matrix is a fluid called blood **plasma**. Plasma consists mainly of water, with a variety of dissolved substances such as nutrients, wastes, enzymes, gases, and hormones. Blood also contains:

- **Red blood cells (erythrocytes):** These transport oxygen to body cells and remove carbon dioxide from them

- **White blood cells (leucocytes):** These are involved in phagocytosis (the destruction of microbes, cell debris, and foreign matter), immunity, and allergic reactions

- **Platelets:** These function in blood clotting

Table 2.9: Classification of muscle tissue

	Skeletal muscle tissue	
	Description	Skeletal muscle tissue is attached to bones and composed of long, cylindrical fibers with multiple nuclei
		The fibers are striated, which means that when looked at under a microscope alternating light and dark bands are seen
		Skeletal muscles are under conscious control and are therefore called voluntary muscles
	Location example	Attached to bones by tendons
	Function	Motion, posture, and heat production
	Cardiac muscle tissue	
	Description	Cardiac muscle tissue forms most of the wall of the heart and consists of branched, striated fibers
		Cardiac tissue is not under conscious control; it is regulated by its own pacemaker and the **autonomic nervous system** and is therefore an involuntary muscle
	Location example	Heart wall
	Function	Pumps blood
	Smooth (visceral) muscle tissue	
	Description	Smooth muscle tissue is so called because it contains non-striated (smooth) fibers
		It is not under conscious control; it is regulated by the autonomic nervous system and is therefore an involuntary muscle
	Location example	Walls of hollow structures, such as blood vessels and intestines
	Function	Constricts and dilates structures to move substances within the body (e.g., blood in blood vessels, foods through GI tract)

Table 2.10: Classification of membranes

Membranes are thin, flexible sheets made up of different tissue layers. They cover surfaces, line body cavities, and form protective sheets around organs. They are categorized as epithelial and connective tissue (**synovial**) membranes

Epithelial membranes: Epithelial membranes consist of an epithelial layer and an underlying connective tissue layer, and include **mucous**, **serous**, and **cutaneous** membranes:

	Mucous membranes	
	Description	Mucous membranes line body cavities that open directly to the exterior
		They are "wet" membranes whose cells secrete mucus, which prevents cavities from drying out, lubricates food as it moves in the gastrointestinal tract, and traps particles in the respiratory tract (note: urine, not mucus, moistens the urinary tract)
		Mucous membranes also act as a barrier that is difficult for pathogens to penetrate
	Location example	The hollow organs of the digestive, respiratory, urinary, and reproductive systems
	Function	Lubrication, movement, and protection
	Serous membranes	
	Description	Serous membranes line body cavities that do not open directly to the exterior, and they cover the organs that lie within those cavities
		They are composed of two layers: the **parietal layer** is attached to the cavity wall, and the **visceral layer** is attached to the organs inside the cavity; between these two layers is a watery lubricating fluid called the **serous fluid**, which enables the organs to move and glide against each other easily
		The layers of a serous membrane consist of thin layers of areolar connective tissue covered by a layer of simple squamous epithelium

Table 2.10: (continued)

	Location example	There are three serous membranes in the body: • **Pleura** lines the thoracic cavity and covers the lungs • **Pericardium** lines the cardiac cavity and covers the heart • **Peritoneum** lines the abdominal cavity and covers the abdominal organs and some of the pelvic organs
	Function	Lubrication and protection
	Cutaneous membrane	
	Description	This is the skin, and it is composed of a superficial keratinizing stratified squamous epithelium and an underlying dense connective tissue layer; it will be discussed in more detail in the following chapter

Connective tissue membranes (synovial membranes): Connective tissue membranes are known as synovial membranes and are composed of areolar connective tissue with elastic fibers and fat. They line:

Bone
Ligament
Hyaline cartilage
Synovial fluid
Joint capsule
Synovial membrane
Periosteum

• The cavities of freely movable joints, and secrete **synovial fluid**, which lubricates the joints
• Bursae and tendon sheaths, which provide cushioning

System Level of Organization of the Body

Tissues combine into structures called **organs**. Organs, for example the heart, have recognizable shapes and specific functions. Organs that share a common function combine to form systems. For example, the digestive system is composed of different organs including the stomach, liver, and pancreas, and its function is to digest food. The systems of the body are the:

- Integumentary system (the skin, hair and nails)
- Skeletal system
- Muscular system
- Nervous system
- Endocrine system
- Respiratory system
- Cardiovascular system
- Lymphatic and immune system
- Digestive system
- Urinary system
- Reproductive system.

These systems work together to ensure that the body maintains a stable internal environment despite any changes in its external environment. This process is known as **homeostasis**.

There are three main approaches to studying anatomy: regional or topographical anatomy which studies the body as parts or segments; clinical or applied anatomy which focuses on the body in relation to the practice of medicine; and systemic anatomy which studies the body's organ systems. This book takes a systemic approach to anatomy.

Integumentary system – *protects the body; regulates body temperature; eliminates waste; helps produce vitamin D; contains nerve endings sensitive to pain, temperature and touch*

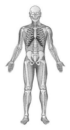

Skeletal system – *provides movement, support and protection; stores minerals; houses the cells that create blood cells*

Muscular system – *powers movement; maintains posture; generates heat*

Nervous system – *regulates body activities through detecting, processing and responding to change in both the external and internal environments*

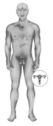

Endocrine system – *regulates body activities through hormones*

Respiratory system – *supplies oxygen and removes carbon dioxide; helps produce vocal sounds*

Cardiovascular system – *transports blood, which carries oxygen, carbon dioxide, nutrients and waste to and from cells; regulates body temperature*

Lymphatic and immune system – *functions in immunity, protection and waste removal; returns proteins and plasma to the cardiovascular system; transports fats*

Digestive system – *breaks down food and absorbs nutrients; eliminates waste*

Urinary system – *eliminates waste; regulates water, electrolyte and acid–base balance of blood*

Reproductive system – *reproduces life*

Theory in practice

It is estimated that we are each made up of about 37.2 trillion cells, and every human cell, except for mature red blood cells, is potentially capable of forming a complete human being. Some cells divide more rapidly than others, and the body naturally loses millions of cells every day; for example, the outer layer of skin is replaced every few days.

Cancer is the uncontrolled division of cells, and common treatments for cancer target quickly dividing cells. Radiation kills cells that divide rapidly, and chemotherapy interferes with a cell's ability to divide. Unfortunately, these therapies can also destroy other quickly dividing cells such as those of the skin, the lining of the mouth, the bone marrow, the hair follicles, and the digestive system. This is why side-effects of these therapies can sometimes include hair loss, mouth ulcers, loss of appetite, nausea, and skin changes.

CHAPTER 3

The Skin, Hair, and Nails

Skin

The skin is the largest organ in the body and combines with the hair and nails to form the integumentary system. It is the only solid protection we have against the external environment, and it reflects the state of our health and emotions.

The skin helps regulate the body's internal temperature; contains sensory receptors sensitive to touch, temperature, pressure, and pain; absorbs, excretes, and secretes different substances; provides a physical barrier to protect against trauma, bacteria, dehydration, ultraviolet radiation, and chemical and thermal damage; and also functions in the synthesis of vitamin D. In addition, the skin contains immune cells and acts as a blood reservoir.

Dermatology is the study of the skin, which is a cutaneous membrane made of two distinct layers:

- The **epidermis** is a tough, waterproof outer layer that is continuously being worn away.
- The **dermis** lies beneath it and is a thicker layer that contains nerves, blood vessels, sweat glands, and hair roots.

The dermis is attached to the **subcutaneous layer** (also called subcutis, superficial fascia, or hypodermis), which anchors the skin to the other organs in the body and provides shock absorption and insulation.

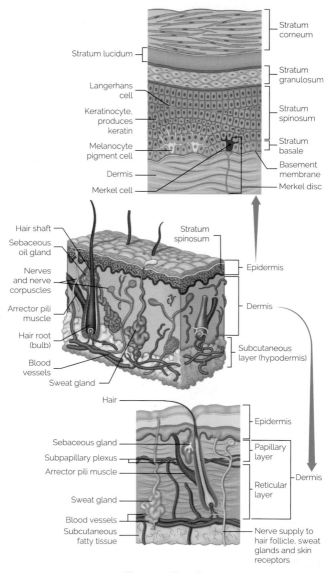

Figure 3.1: The skin

Table 3.1: Tissues that make up the skin

Name	Tissue Type	General Description
Epidermis		
Epidermis	Keratinized stratified squamous epithelium	A thin layer of flat, dead cells that are continually being shed; this is a waterproof and protective layer
Four types of cells are found in the epidermis:		
Keratinocytes	Produce keratin, a protein that helps waterproof and protect the skin	
Melanocytes	Produce melanin, a pigment that contributes to skin color and absorbs ultraviolet light	
Langerhans cells	Respond to foreign bodies and thus play a role in skin immunity	
Merkel cells	Make contact with nerve cells to form Merkel disks, which function in the sensation of touch. They are found in the stratum basale of hairless skin and are attached to keratinocytes	
The epidermis is composed of five layers of stratified squamous epithelial tissue, which becomes tough and hard through a process called **keratinization**. From the deepest to the most superficial layer, they are the:		
Stratum basale/stratum germinativum (basal-cell layer)	The deepest layer of the epidermis and the base from where new cells germinate	
Stratum spinosum (prickle-cell layer)	Consists of prickly cells that are beginning to go through the process of keratinization	
Stratum granulosum (granular-cell layer)	Composed of degenerating cells becoming increasingly filled with granules of keratin	
Stratum lucidum (clear-cell layer)	A waterproof layer of dead, clear cells	
Stratum corneum (horny-cell layer)	The outermost layer of the skin. Its cells are dead and filled with keratin. Thus, they are tough, durable, and horny. The cells are constantly being shed through the process of desquamation	
The epidermis can also be divided into two major layers:		

Table 3.1: (*continued*)

Name	Tissue Type	General Description
Epidermis (*continued*)		
Inner layer (Malpighian layer)	Contains only dividing, non-keratinized cells. It includes the stratum basale, where new cells are constantly being produced, and the stratum spinosum, where some cells are only beginning to go through the process of keratinization	
Outer layer	Composed of keratinized, non-dividing cells. It includes the strata granulosum, lucidum, and corneum	
Dermis		
Dermis	Connective tissue containing collagen and elastic fibers	A thicker layer that supports the epidermis above it by enabling the passage of nutrients and oxygen; it also allows the skin to move, absorbs shocks, and cools and warms the body
The dermis is the supportive layer beneath the epidermis and is composed of connective tissue that contains both collagen and elastic fibers. The dermis can be divided into two layers:		
Papillary layer	This superficial layer has finger-like projections that go into the epidermis. These projections contain loops of capillaries and nerve endings sensitive to touch	
Reticular layer	This deeper layer houses hair follicles, nerves, oil glands, ducts of sweat glands, and adipose tissue	

Types of Skin

Now that you know the structure and functions of the skin, it is time to take a look at what you really see on a person and what distinguishes one person's skin from another. The common skin types are:

- **Normal (balanced) skin:** Normal skin is a balanced skin in which there are no signs of oily or dry areas. It is actually a "perfect" skin and is, of course, quite rare in adults. Normal skin:
  - Has an even texture
  - Has good elasticity
  - Has small pores
  - Feels soft and firm to the touch
  - Is usually blemish-free

- **Oily skin:** In an oily skin there is an overproduction of sebum by the sebaceous glands. This can be caused by hormones, e.g., in puberty. Oily skin:
    - Has an uneven texture
    - Has normal elasticity
    - Has large pores
    - Feels thick and greasy to touch
    - Often has blemishes such as comedones, papules, pustules, and scars
    - Appears sallow (unhealthy yellow or pale brown in color) and has a characteristic shine
    - Ages slowly
- **Dry skin:** In dry skin there is either an underproduction of sebum or a lack of moisture, or both together. Dry skin:
    - Has flaky, dry patches and a thin, coarse texture
    - Has poor elasticity
    - Feels dry, coarse, and papery to touch
    - Looks like parchment and often has dilated capillaries around the cheek and nose areas
    - May also be sensitive
    - Ages prematurely, especially around the eyes, mouth and neck
- **Combination skin:** Combination skin is a mixture of dry, normal, and greasy skin and is the most common type of skin. Combination skin:
    - Has an oily T-zone (forehead, nose, and chin), which feels thick and greasy and is often blemished
    - Has dry cheeks and neck, which feel flaky and coarse and may have dilated capillaries
    - Has varying tone and elasticity
    - May also have sensitive areas
- **Sensitive skin:** Sensitive skin often accompanies dry skin and is easily irritated. Sensitive skin:
    - Is hypersensitive and reactions can include reddening, itching, and chafing
    - Often has dilated capillaries
    - Is often dry and transparent
    - Is usually warm to touch
- **Mature skin:** As a person ages, both sun exposure and hormonal changes affect the skin, causing it to age. Mature skin:
    - Is dry, thin, less elastic, and bruises easily
    - Can become sensitive and irritated and take longer to heal
    - Can have pigmentation, deep wrinkles, and dilated capillaries

Hair

Hairs (or pili) are columns of keratinized dead cells that function in helping to protect the body and maintain its temperature. They grow over most of the body, except the palms of the hands, the soles of the feet, the eyelids, lips, and nipples.

Hair grows out of hair follicles, which surround the root of the hair. At the base of the follicle is the bulb, which houses the papilla containing areolar and connective tissue, blood vessels, and the hair matrix. Living cells in the matrix divide constantly and push the hair upward.

As the cells move upward, they fill with keratin and die, forming the hair shaft. Sebaceous oil glands secrete sebum, which lubricates the hair. Arrector pili muscles are bands of smooth muscle cells that attach the side of the hair to the dermis and contract to pull the hairs into a vertical position as a response to cold, fright, or differing emotions.

Table 3.2: Types of hair

Hair Type	Description and Location
Lanugo	Soft hair that begins to cover a fetus from the third month of pregnancy. It is usually shed by the eighth month of pregnancy. Found only on a fetus
Vellus	Soft and downy hair. Found all over the body, except the palms of the hands, soles of the feet, eyelids, lips, and nipples
Terminal	Longer, coarser hair. Found on the head, eyebrows, eyelashes, under the arms and in the pubic area

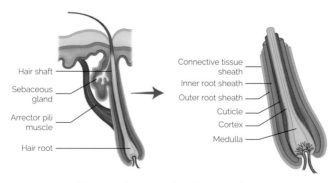

Figure 3.2: Structure of a hair and hair follicle

Life cycle of a hair

The life cycle of a hair includes growing, transitional, and resting stages.

The growth stage is called **anagen**, and in this stage a follicle reforms and the matrix divides to create a new hair. As with the skin, the older cells are pushed upward as new cells develop below. The transitional stage is known as **catagen** and usually lasts one to two weeks. During this stage the hair separates from the base of the follicle as the dermal papilla breaks down. Finally, in the resting stage, or **telogen**, the follicle is no longer attached to the dermal papilla and the hair moves up the follicle and is naturally shed.

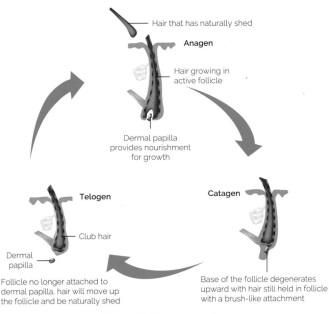

Hair that has naturally shed

Anagen

Hair growing in active follicle

Dermal papilla provides nourishment for growth

Telogen

Club hair

Dermal papilla

Follicle no longer attached to dermal papilla, hair will move up the follicle and be naturally shed

Catagen

Base of the follicle degenerates upward with hair still held in follicle with a brush-like attachment

Figure 3.3: Life cycle of a hair

Nails

Nails are hard plates that cover and protect the ends of the fingers and toes and enable us to pick up small objects easily. Beneath the nail plate is the germinal matrix where cell division and nail growth occur. Living cells at the matrix divide constantly and push the nail forward. As the cells move toward the fingertips, they fill with tough keratin and die, forming the hard nail plate.

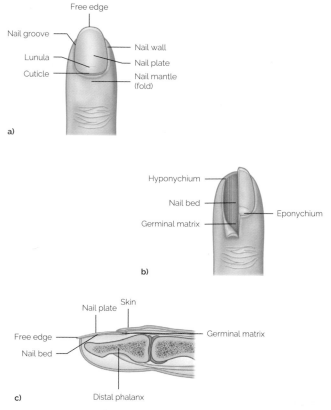

Figure 3.4: Structure of the nail

Cutaneous Glands

Table 3.3: Cutaneous glands

Gland	Product	Function
Sebaceous (oil) glands	Secrete sebum	Sebum acts as a lubricant to prevent hair from drying out. It also prevents excessive evaporation of moisture from the skin and inhibits the growth of certain bacteria
Sudoriferous (sweat) glands	Excrete sweat onto the surface of the skin	Aids heat regulation of the body, helps eliminate small amounts of waste, and helps inhibit the growth of bacteria on the skin's surface

Sudoriferous glands are divided into two types:

	Eccrine Glands	Apocrine Glands
Location	These are very common sweat glands found all over the body except for on the lips, nail beds, some of the reproductive organs, and the eardrums; they are most numerous on the palms of the hands and the soles of the feet	These are found in the axillae (armpits), pubic region, and the areolae of the breasts
Structure	Their secretory portion is located in the subcutaneous layer and their duct extends outward through the skin, opening as a pore on the surface of the skin	Their secretory portion is located in the dermis or subcutaneous layer and their duct opens into hair follicles
Excretion	Sweat	Sweat, fatty acids, and proteins
Function	They excrete waste and help to regulate the body's temperature by keeping it cool	Their exact function is not yet known, but they are activated by pain, stress, and sexual foreplay

Theory in practice

Do you ever wonder what determines the color of your skin? All races have approximately the same number of melanocytes in their skin, yet differing amounts of pigments, and it is these pigments that make a difference to skin color. The three pigments in your skin that are responsible for your color are *melanin*, which varies in color from pale yellow to black; *carotene*, which is a yellowish-orange pigment; and *hemoglobin*, which is the pigment that carries oxygen in the red blood cells.

Your hair color is also determined by differing amounts of pigments. For example, people with dark hair have mainly true melanin in their hair; blondes and redheads have melanin and varying amounts of the minerals iron and sulfur; and those with gray hair have a decreased amount of the enzyme needed for the synthesis of melanin.

How does aging affect our skin? We are born with soft, smooth skin that has a thick layer of fat and only a thin layer of protective keratin. It is not a very effective barrier against harmful substances. As we grow our skin thickens and strengthens, and by adulthood we have strong, supple skin that functions effectively in protecting us and regulating our body temperature.

However, as we age, the skin begins to thin. In old age both the dermis and epidermis are thinner, and a lot of the underlying fat layer that usually insulates us has disappeared. The number of sweat glands and blood vessels in our skin also decreases, and this lowers our skin's ability to regulate our body temperature. Therefore, we become more susceptible to both cold and heat. The skin also loses its elasticity with old age and begins to wrinkle and sag, and there is a decrease in melanocytes, which causes an increase in sensitivity to the sun. Finally, a decrease in the number of nerve endings results in less sensitivity to external stimuli.

CHAPTER 4

The Skeletal System

Bones

We are born with approximately 350 bones, but gradually some of these fuse together until puberty, when we have only 206 bones. These bones form the supporting structure of the body, and are collectively known as the *endoskeleton*. (The *exoskeleton* is well developed in many invertebrates, but exists in humans only as teeth, nails, and hair.)

Fully developed bone is the hardest tissue in the body and is composed of 20% water, 30% to 40% organic matter, and 40% to 50% inorganic matter. Bones support and protect the body, allow for movement and mineral homeostasis, and are a site of blood-cell production as well as energy storage.

Bone tissue

Osteology is the study of bone. **Osseous tissue** (bone tissue) is a connective tissue whose matrix is composed of water, protein, fibers, and mineral salts. The fibers are made of a protein called **collagen**, which enables bone to resist being stretched or torn apart. This is known as "tensile strength," and without collagen bones would be hard and brittle. The mineral salts are mainly calcium carbonate and a crystallized compound called **hydroxyapatite**. These salts give bone its hardness.

The cells that make up bone tissue are:

- **Osteoprogenitor cells:** These are stem cells derived from mesenchyme (the connective tissue found in an embryo). They have the ability to become osteoblasts.
- **Osteoblasts:** These cells secrete collagen and other organic components to form bone.

- **Osteocytes:** These are mature bone cells that maintain the daily activities of bone tissue. They are derived from osteoblasts and are the main cells found in bone tissue.
- **Osteoclasts:** These cells are found on the surface of bones, and they destroy or resorb bone tissue.

Two types of bone tissue exist: **compact** and **spongy**.

Compact (dense) bone tissue

This is a very hard, compact tissue that has few spaces within it. It is composed of a basic structural unit called an **osteon** or **Haversian system**, which consists of concentric rings called **lamellae** made of a hard, calcified matrix. Between the lamellae are small spaces called **lacunae** where osteocytes are housed. Through the center of the lamellae run **Haversian canals**, in which nerves and blood and lymph vessels are found. These Haversian canals are connected to one another and the periosteum through perforating channels called **Volkmann's canals**. **Canaliculi** are tiny canals that radiate outward from the central canals to other lacunae.

The main functions of compact bone tissue are protection and support. It forms the external layer of all bones.

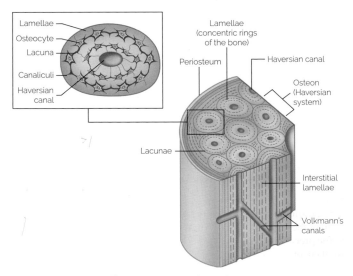

Figure 4.1: Compact bone tissue

Spongy (cancellous) bone tissue

This is a light tissue with many spaces within it, and it has a sponge-like appearance. It does not contain osteons. Instead, it is made up of lamellae arranged in an irregular latticework of thin plates of bone called **trabeculae**. Within the trabeculae are lacunae containing osteocytes. Spongy bone tissue contains red bone marrow, which is the site of blood-cell production. It is found in the hip bones, ribs, sternum, vertebrae, skull, and the ends of some long bones.

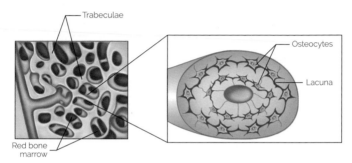

Figure 4.2: Spongy bone tissue

Bone formation and remodeling

Bone is a dynamic, living tissue that is constantly changing, repairing, and reshaping itself. Most bones are formed through **ossification**, a process that begins somewhere between the sixth and seventh week of embryonic life and continues throughout adulthood. There are two types of ossification:

- **Intramembranous ossification:** Where bone forms on or within loose, fibrous connective-tissue membranes without first going through a **cartilage** stage.
- **Endochondral ossification:** Where bone forms on **hyaline cartilage,** which has been produced by cells called **chondroblasts**.

New bone tissue constantly replaces old, worn-out, or injured bone tissue through the process of **remodeling**. Mechanical stress, in the form of the pull of gravity and the pull of skeletal muscles, is integral to the process of remodeling. If these stresses are absent the bones weaken, or if they are excessive the bones thicken abnormally.

Types of Bones

Bones are usually classified into five types according to their features, such as shape, placement, and additional properties. Types of bones include long, short, flat, irregular, and sesamoid.

Table 4.1: Types of bone

Classification	Description	Example
Long	Have a greater length than width and usually contain a longer shaft with two ends. They are slightly curved to provide strength and are composed of mainly compact bone tissue with some spongy bone tissue	Femur, tibia, fibula, phalanges, humerus, ulna, and radius
Short	Cube-shaped and nearly equal in length and width. They are made up of mainly spongy bone with a thin surface of compact bone	Carpals and tarsals
Flat	Thin bones consisting of a layer of spongy bone enclosed by layers of compact bone. Flat bones act as areas of attachment for skeletal muscles, and also provide protection	Cranial bones, sternum, ribs, and scapulae
Irregular	Most bones that cannot be classified as long, short, or flat bones fall into the category of irregular bones. They have complex shapes and varying amounts of compact and spongy tissues	Vertebrae
Sesamoid	Oval bones that develop in tendons where there is considerable pressure	Patella (kneecap)

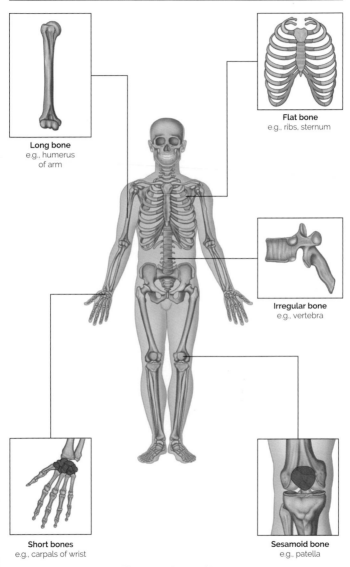

Long bone
e.g., humerus
of arm

Flat bone
e.g., ribs, sternum

Irregular bone
e.g., vertebra

Short bones
e.g., carpals of wrist

Sesamoid bone
e.g., patella

Figure 4.3: Types of bone

Structure of a long bone

A long bone has a main, central shaft called a **diaphysis**. The diaphysis is covered by a membrane known as the **periosteum**. The periosteum provides attachment for muscles, tendons, and ligaments and is also essential for nutrition, repair, and growth in diameter of bone.

The periosteum consists of two layers:

- **Outer fibrous layer:** Made of dense irregular connective tissue that contains blood vessels, lymph vessels, and nerves that pass into the bone
- **Inner osteogenic layer:** Made of elastic fibers and containing blood vessels and bone cells.

Each end of the diaphysis is called an **epiphysis**. Each epiphysis is covered by a thin layer of hyaline cartilage called **articular cartilage**. This cartilage reduces friction and absorbs shock at the area where the bone forms an articulation (joint) with the surface of another bone. The epiphysis is made of mainly spongy bone tissue and contains red bone marrow. This is where blood cells are produced.

The region where the diaphysis joins the epiphysis is the **metaphysis**. In a growing bone the metaphysis has a layer of hyaline cartilage that allows the diaphysis to grow in length. This is called the **epiphyseal plate**. In a mature bone that is no longer growing in length, the epiphyseal plate is replaced by the **epiphyseal line**.

Inside the diaphysis is a space known as the **medullary** or **marrow cavity**. The medullary cavity is lined by a membrane called the **endosteum**. This contains cells necessary for bone formation. In adults this cavity contains fatty yellow bone marrow, which stores lipids.

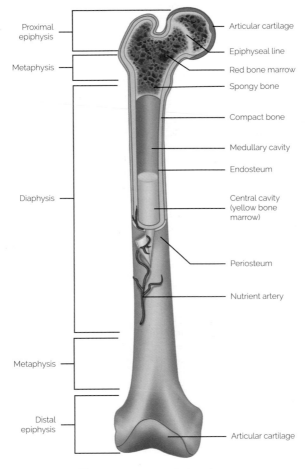

Figure 4.4: Structure of a long bone

Organization of the Skeleton

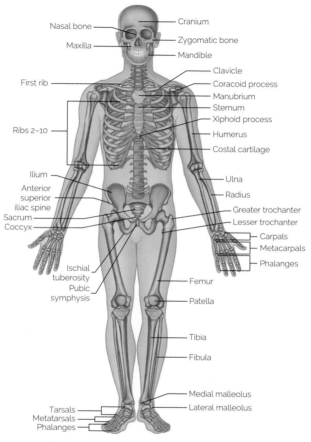

Figure 4.5: Overview of the skeleton (anterior view)

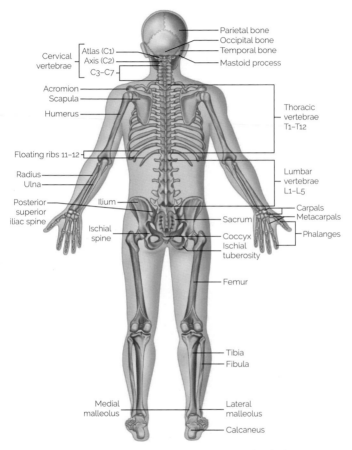

Figure 4.6: Overview of the skeleton (posterior view)

Axial Skeleton

Bones of the skull

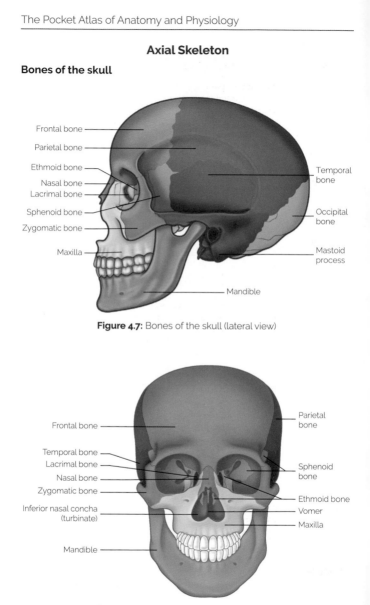

Figure 4.7: Bones of the skull (lateral view)

Figure 4.8: Bones of the skull (anterior view)

Table 4.2: Bones of the skull

Parietal	Two parietal bones form the sides and roof of the cranium
Temporal	Beneath the parietal bones are two temporal bones; they form the inferior lateral sides of the cranium and part of the cranial floor
Frontal	Forms the forehead and the roofs of the orbits (eye sockets)
Occipital	Forms the back of the cranium and most of the base of the cranium
Sphenoid	This butterfly-shaped bone articulates with all the other cranial bones and holds them together; it lies at the middle part of the base of the skull and forms part of the floor of the cranium, the sides of the cranium and parts of the eye orbits
Ethmoid	This is the major supporting structure of the nasal cavity; it forms the roof of the nasal cavity and part of the medial walls of the eye orbits
Nasal	These two bones form the bridge of the nose
Maxillae	These two bones unite to form the upper jawbone, part of the floors of the orbits, part of the lateral walls and floor of the nasal cavity, and most of the roof of the mouth; the maxillae articulate with every bone in the face except the mandible
Zygomatic	These two bones are the cheekbones and form part of the lateral wall and floor of the orbits
Lacrimal	These tiny bones form part of the medial wall of the eye orbit; they are about the same size and shape as a fingernail and are the smallest bones in the face
Palatine	These two bones form part of the palate (roof of the mouth), part of the floor and lateral wall of the nasal cavity, and part of the floors of the orbits
Inferior nasal conchae (turbinates)	These two bones form part of the lateral wall of the nasal cavity and help to circulate, filter, and warm air before it passes into the lungs
Vomer	This bone forms part of the nasal septum, which divides the nose into left and right sides
Mandible	This is the lower jaw; it is the only moveable bone in the skull and is also the largest and strongest facial bone

Bones of the neck and spine

The neck comprises the cervical vertebrae and the hyoid bone. The hyoid bone is suspended from the temporal bones by ligaments and muscles. It supports the tongue and provides attachment for some of the muscles of the neck and pharynx. Please refer to Figure 8.5 on page 196 for an image of the hyoid bone.

The spine is composed of 33 vertebrae: 7 cervical, 12 thoracic, 5 lumbar, 5 sacral, and 4 coccygeal. The sacral vertebrae fuse to form the sacrum, and the coccygeal vertebrae fuse to form the coccyx (tailbone). Thus, there are 26 separate bones that make up the spine.

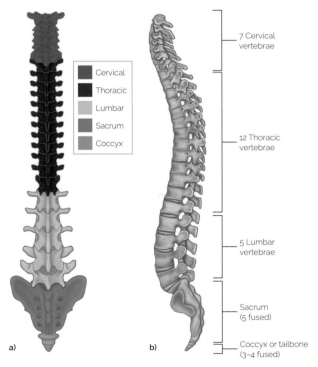

Figure 4.9: Vertebral column: (a) posterior view, (b) lateral view

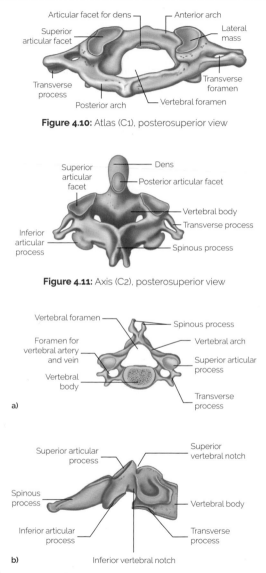

Figure 4.10: Atlas (C1), posterosuperior view

Figure 4.11: Axis (C2), posterosuperior view

Figure 4.12: Cervical vertebra (C5), (a) superior view; (b) lateral view

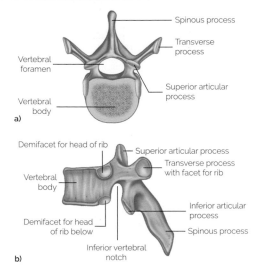

Figure 4.13: Thoracic vertebra (T6), (a) superior view; (b) lateral view

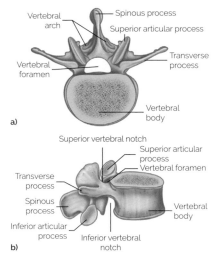

Figure 4.14: Lumbar vertebra (C5), (a) superior view; (b) lateral view

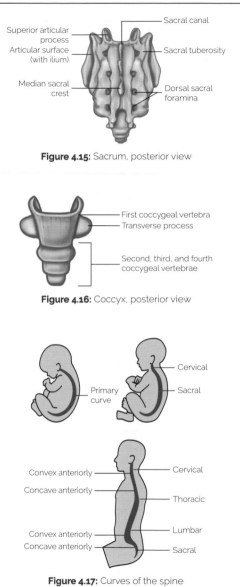

Figure 4.15: Sacrum, posterior view

Superior articular process
Articular surface (with ilium)
Median sacral crest
Sacral canal
Sacral tuberosity
Dorsal sacral foramina

Figure 4.16: Coccyx, posterior view

First coccygeal vertebra
Transverse process
Second, third, and fourth coccygeal vertebrae

Figure 4.17: Curves of the spine

Primary curve
Cervical
Sacral

Convex anteriorly — Cervical
Concave anteriorly — Thoracic
Convex anteriorly — Lumbar
Concave anteriorly — Sacral

Bones of the thorax

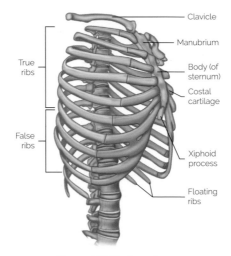

Figure 4.18: The bony thorax

Table 4.3: Bones of the thorax

Sternum	This is the breastbone; it is a flat, narrow, long bone that is found in the middle of the anterior thoracic wall. The sternum is composed of three parts: • The superior **manubrium**, which articulates with the clavicle and the costal cartilages of the first two ribs • The long, middle **body**, which articulates with the costal cartilages of the second through seventh ribs • The inferior **xiphoid process**, which provides attachment for some abdominal muscles
Ribs	There are 12 pairs of ribs: • The first 7 pairs of ribs are attached to the sternum by a type of hyaline cartilage called **costal cartilage**; these 7 pairs are known as **true ribs** because they are directly attached to the sternum

Table 4.3: (continued)

	• The remaining 5 pairs of ribs are **false ribs** because they are not directly attached to the sternum • The 8th, 9th, and 10th pairs are attached to each other by their cartilages and then to the cartilages of the 7th pair of ribs • The 11th and 12th pairs of ribs are **floating ribs** that are only attached to abdominal muscles
Thoracic vertebrae	See section "Bones of the neck and spine" above for more information on the thoracic vertebrae

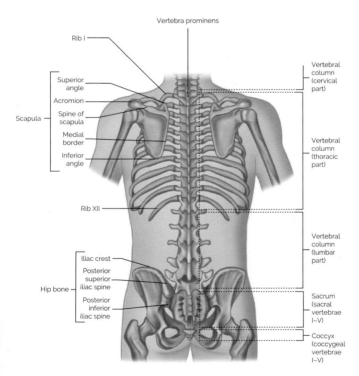

Figure 4.19: Bones of the trunk and pectoral and pelvic girdles (posterior view)

Appendicular Skeleton

Bones of the upper limb and shoulder girdle

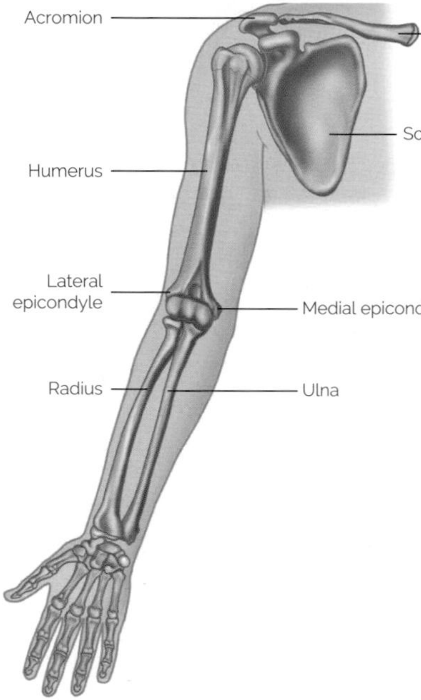

Figure 4.20: Bones of the upper limb and shoulder girdle

Table 4.4: Bones of the upper limb and shoulder girdle

Shoulder, Arm, and Forearm	
Clavicle	This is the collarbone and it is a long, slender, double-curved bone that helps hold the arm away from the top of the thorax; it also helps prevent shoulder dislocation
Scapula	This is the shoulder blade; it is a large, flat, triangular bone
Humerus	This is the longest and largest bone of the upper limb and is the arm bone
Ulna	This bone is located on the medial aspect (little-finger side) of the forearm when the body is in the anatomical position; its proximal end is called the **olecranon**, which is commonly known as the elbow
Radius	This is located on the lateral aspect of the forearm (thumb side)

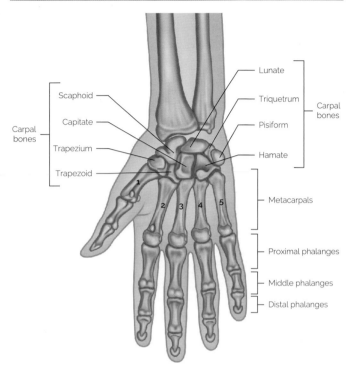

Figure 4.21: Bones of the wrist and hand

Table 4.5: Bones of the wrist and hand

Wrist and Hand	
Carpals (8)	The wrist consists of 8 small bones arranged in two irregular rows of 4 bones each, bound together by ligaments These bones are the: **trapezium, trapezoid, capitate, hamate, scaphoid, lunate, triquetrum,** and **pisiform**
Metacarpals (5)	The palm of the hand is formed of 5 metacarpals; they are numbered 1 to 5, starting with the thumb side of the hand
Phalanges (14)	The fingers are made up of 14 phalanges; in each finger are a proximal, middle, and distal phalange (phalanx), but in the thumb there are only a proximal and distal phalange; the thumb is sometimes called the **pollex**

Table 4.6: Bones of the pelvic girdle, thigh, and leg

Pelvic Girdle	
Ilium	This large, winglike bone forms the superior portion of the hip bone; its upper border serves as a site of attachment for many muscles and is called the **iliac crest**
Ischium	This forms the inferior and posterior portion of the hip bone
Pubis	This is the most anterior part of the hip bone
Thigh and Knee	
Femur	This is the thigh bone and it is the longest, strongest, and heaviest bone in the body
Patella	This is the kneecap and it is a sesamoid bone that is attached to the tibia by the patellar ligament; it develops in the tendon of the quadriceps femoris muscle to protect the knee joint and help maintain the position of the tendon when the knee is bent
Leg	
Tibia	This is the shinbone and is the large, medial bone of the leg
Fibula	This is a thin bone that runs parallel to the tibia

Bones of the pelvic girdle

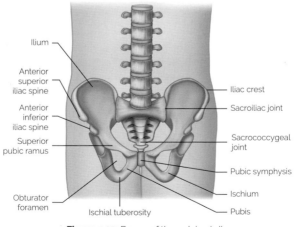

Figure 4.22: Bones of the pelvic girdle

Bones of the lower limb

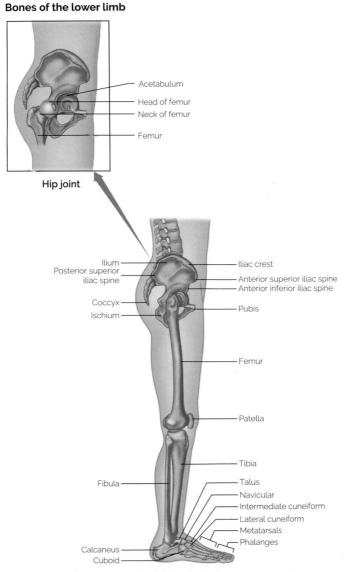

Acetabulum
Head of femur
Neck of femur
Femur

Hip joint

Ilium
Posterior superior iliac spine
Coccyx
Ischium

Iliac crest
Anterior superior iliac spine
Anterior inferior iliac spine
Pubis

Femur

Patella

Tibia

Fibula

Talus
Navicular
Intermediate cuneiform
Lateral cuneiform
Metatarsals
Phalanges

Calcaneus
Cuboid

Figure 4.23: Lateral view of the pelvic girdle, lower limb, and foot

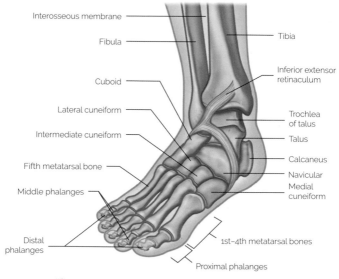

Figure 4.24: Bones of the right foot, anteromedial view

Table 4.7: Bones of the foot

Foot	
Tarsals (7)	The **tarsus**, or back portion, of the foot is made up of 7 tarsal bones: the **talus** (ankle bone), **calcaneus** (heel bone), **cuboid**, **navicular**, and three **cuneiforms** (medial, intermediate, and lateral)
Metatarsals (5)	The **metatarsus** of the foot is made up of 5 metatarsals, numbered 1 to 5, starting with the large-toe (medial) side of the foot
Phalanges (14)	The toes are made up of 14 phalanges. In each toe are a proximal, middle, and distal phalange (phalanx), but in the large toe (big toe) there are only a proximal and distal phalange; the large toe is sometimes called the **hallux**

The arches of the foot

The bones of the foot arrange themselves into three distinct arches: the lateral longitudinal, the medial longitudinal, and the transverse. Their shape allows them to act like a spring, bearing the weight of the body, as well as absorbing shock produced by movement. The flexibility of these arches facilitates functions such as walking and running.

The medial longitudinal arch is made up of the talus, calcaneus, navicular, three cuneiforms, and first three metatarsals. This arch bears most of the body's weight. Pes cavus is a foot condition characterised by an unusually high medial longitudinal arch, and because of the higher arch, there is a diminished ability to shock absorb during walking with an increased degree of stress placed on the ball and heel of the foot.

The lateral longitudinal arch is made up of four bones: the calcaneus, cuboid, and fourth and fifth metatarsals. Pes planus (flat footed) is a common foot condition in which the longitudinal arches have been lost.

The transverse arch is located on the coronal plane, and comprises the three cuneiforms, the cuboid, and the metatarsal bases.

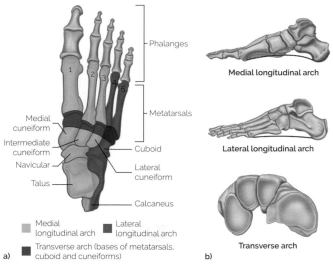

Figure 4.25: Arches of the foot

Joints

Joints are also called **articulations** and are the points of contact between bones, cartilage and bones, or teeth and bones. They function not only in holding the bones together but also in providing movement and helping protect organs. **Arthrology** is the study of joints.

Table 4.8: Functional classification of joints

Functional Classification	Description	Example
Synarthroses	Immovable joints that do not allow movement	Coronal suture
Amphiarthroses	Slightly movable joints that permit a minimal amount of flexibility and movement	Pubic symphysis
Diarthroses	Freely movable joints that contain a synovial cavity and permit a number of different movements	Glenohumeral joint

Table 4.9: Structural classification of joints

Structural Classification	Description	Example
Fibrous	Bone ends are held together by fibrous (collagenous) connective tissue, and there is no synovial cavity between them. Thus, they are strong joints that do not permit movement. In general, they are synarthrotic	Sutures
Cartilaginous	Bone ends are held together by cartilage, and they do not have a synovial cavity between them. Thus, these are also strong joints that permit only minimal movement. Most cartilaginous joints are amphiarthrotic, although some can be synarthrotic	Pubic symphysis
Synovial	Bone ends are separated by a synovial cavity containing synovial fluid. This allows for a great deal of movement; all synovial joints are diarthrotic	Tibiofemoral joint

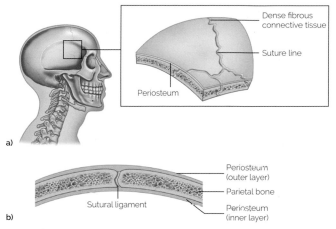

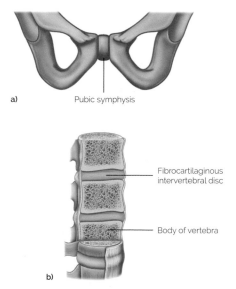

Figure 4.26: Fibrous joint, as shown by the sutures of the skull

Figure 4.27: Cartilaginous joint, as shown by (a) the pubic symphysis; (b) intervertebral joints

Synovial joints

Synovial joints predominate in the limbs of the body. They are enclosed in a synovial (articular) capsule composed of an outer fibrous capsule and an inner synovial membrane that secretes synovial fluid. This fluid fills the synovial cavity which is the space between the articulating bones.

The surfaces of the bones are covered in hyaline (articular) cartilage and the entire synovial capsule is reinforced with ligaments. Together these features allow for a great deal of movement. Some synovial joints also have articular discs (menisci) which are pads of fibrocartilage lying between the articular surfaces of the bones, and bursae which are sac-like structures made of connective tissue and filled with synovial fluid.

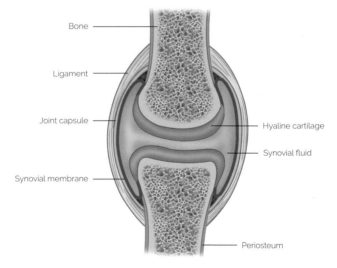

Figure 4.28: Synovial joint, as shown by the knee

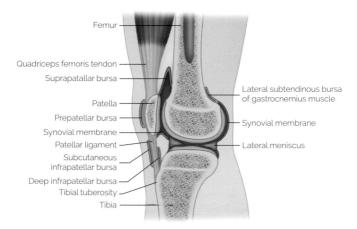

Figure 4.29: Shock-absorbing and friction-reducing structures of a synovial joint.

Table 4.10: Movements at synovial joints

General	
Movement	**Definition**
Flexion	This is the bending of a joint in which the angle between articulating bones decreases
Extension	This usually restores a body part to its anatomical position after it has been flexed, and is the straightening of a joint in which the angle between the articulating bones increases
Hyperextension	This is when a body part extends beyond its anatomical position
Flexion, extension, and hyperextension	Here is a simple exercise to demonstrate the movements of flexion, extension, and hyperextension. Nod your head. When your chin touches your throat, you are *flexing* your neck. When you return your head to its normal upright position, you are *extending* your neck. When you push your head backward so that your chin is facing up toward the ceiling, you are *hyperextending* your neck

Table 4.10: (*continued*)

Abduction	This is a movement away from the midline of the body
Adduction	This is the opposite of abduction and is a movement toward the midline of the body
Circumduction	This is a circular movement of the distal end of a body part, and it involves a succession of flexion–extension and abduction–adduction
  Abduction and adduction Circumduction	These common arm exercises demonstrate abduction, adduction, and circumduction. Do some star jumps. As you swing your arms up and away from your body, you are *abducting* them. Likewise, you are *abducting* your legs as you jump them outward. As you bring your arms back down toward the sides of your body, you are *adducting* them. You are also *adducting* your legs as you jump them together again. Swing your arms in a large circular movement as if you are warming up your shoulders. This circular swinging movement is *circumduction*
Rotation	This is the movement of a bone in a single plane around its longitudinal axis
Medial or internal rotation	Involves the movement of the anterior surface of a bone toward the midline
Lateral or external rotation	Involves the movement of the anterior surface of a bone away from the midline

Table 4.10: (continued)

General	
Movement	**Definition**
 Rotation	Turn your head from side to side as if you are saying "no." You are *rotating* your head. Now, stand in the anatomical position with your palms facing forward and turn your palms in toward your thighs so that they face backward—you have *medially rotated* your forearms. Turn your palms from facing backward to their original position in which they were facing forward—you have *laterally rotated* your forearms

Forearm	
Movement	**Definition**
Pronation	This involves turning your palm posteriorly or inferiorly
Supination	This involves turning your palm anteriorly or superiorly
 Pronation and supination	Stand with your arms out in front of you, with your palms facing the floor. Now turn your palms up toward the ceiling—you are *supinating* your forearm. Now turn them back down toward the floor—you are *pronating* your forearm

Foot	
Movement	**Definition**
Inversion	This is turning the sole of the foot inward
Eversion	This is the opposite of inversion, and is turning the sole of the foot outward
Dorsiflexion	This is the pulling of the foot upward toward the shin, in the direction of the dorsum

Table 4.10: (continued)

Foot (continued)	
Movement	**Definition**
Plantar flexion	This is the opposite of dorsiflexion, and is the pointing of the foot downward, in the direction of the plantar surface
Inversion and eversion Plantar flexion and dorsiflexion	Here are a few foot exercises you can do to demonstrate inversion, eversion, dorsiflexion, and plantar flexion Sit on the floor with your legs out straight. Roll your feet inward so that your soles are facing one another—you are *inverting* them. Now roll your feet away from one another so that your soles are facing away from each other—you are *everting* them Still sitting on the floor with your legs out straight, have your toes pointing up toward the ceiling. Now point your toes forward as if you are trying to touch the floor with them— you are *plantar flexing* your feet Then do the opposite movement, trying to pull your toes upward and back toward your shins—you are *dorsiflexing* them

Table 4.11: Classification of synovial joints

Name of Joint	Shapes of Articulating Surfaces	Movements Permitted	Examples
Gliding (plane)	Flat surfaces meet	Side to side Back and forth **Note:** no angular or rotary motions are permitted	Patellofemoral joint at knee Intercarpal joints Intertarsal joints Sacroiliac joint

Table 4.11: (continued)

Name of Joint	Shapes of Articulating Surfaces	Movements Permitted		Examples
Hinge	A convex surface fits into a concave one	Flexion Extension **Note:** movements are in a single plane only		Elbow joint Tibiofemoral joint at knee Ankle joint IP joints
Pivot	A rounded/pointed surface fits into a ring	Rotation		Atlas and axis Ulna and radius
Condyloid (ellipsoid)	A condyle is a rounded/oval protuberance at the end of a bone, and it fits into an elliptical cavity	Back and forth Flexion Extension Abduction Adduction Circumduction		Wrist joint MCP joints
Saddle	A surface shaped like the legs of a rider fits into a saddle-shaped surface	Side to side Back and forth Flexion Extension	Abduction Adduction Circumduction	Thumb joint
		Opposition of thumbs (where the tip of the thumb crosses the palm and meets the tip of a finger)		
Ball and socket (spheroidal)	A ball fits into a cup-shaped socket	Flexion Extension Abduction Adduction Rotation Circumduction		Shoulder joint Hip joint

Theory in practice

Bones are not dead materials that simply support your body. They are living structures that are constantly changing by reshaping, rebuilding, and repairing themselves.

An embryo initially has no bones, and its "skeleton" is composed entirely of fibrous connective-tissue membranes, called **mesenchyme**, and hyaline cartilage. Around the sixth or seventh week of embryonic life, ossification begins and, although bone growth in length is usually completed around 25 years of age, bones continue to thicken and be remodeled throughout life. The face only stops growing when one reaches about 16 years of age, and fusion of the coccyx takes place somewhere between 20 and 30 years of age. Finally, the xiphoid process of the sternum does not ossify until about the age of 40 years.

Unlike most tissues, which form scar tissue if injured, bones can completely heal themselves without scarring. This is because bone tissue is constantly replacing itself through the process of remodeling. Immediately after being fractured, the damaged area of the bone becomes swollen. This inflammatory response may take several weeks but is vital to clearing the area of any damaged or dead cells. A fibrocartilaginous **callus** is then formed in the area of repair. This is a soft, rubbery tissue, which slowly becomes mineralized and strengthens into a bony callus. Finally, remodeling of the bone takes place, where the external callus is slowly replaced by stronger bone. The entire healing process can take many months.

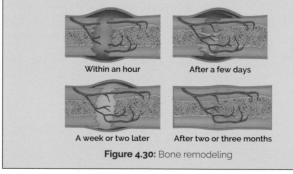

Within an hour · After a few days

A week or two later · After two or three months

Figure 4.30: Bone remodeling

CHAPTER 5

The Muscular System

Muscles have the unique ability of shortening themselves. This is known as **contraction** and is the essential function of all muscles. Through contracting, muscles can produce movement, maintain posture, move substances within the body, regulate organ volume, and produce heat (thermogenesis).

Muscle Tissue

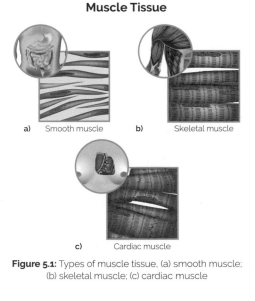

a)　　Smooth muscle　　　　b)　　Skeletal muscle

c)　　Cardiac muscle

Figure 5.1: Types of muscle tissue, (a) smooth muscle; (b) skeletal muscle; (c) cardiac muscle

Table 5.1: Characteristics of the three types of muscle tissue

Type	Anatomy	Location	Function	Control
Skeletal	Muscle fibers: Are striated Are generally long and cylindrical-shaped (although some are circular) Have multiple nuclei and many mitochondria Are strengthened and reinforced by connective tissue Connect to tendons	Attached by tendons to bones, skin, or other muscles	Enables movement of the skeleton Enables movement of lymph and venous blood Maintains posture Produces heat	Voluntary
Cardiac	Muscle fibers: Are striated Are arranged in spiral-shaped bundles of branching cells	Forms most of the heart	Pumps blood around the body Helps regulate blood pressure	Involuntary Contracts at a steady rate set by a "pacemaker" that is adjusted by neurotransmitters and hormones
Smooth (visceral)	Muscle fibers: Are non-striated, hence they are called smooth Are spindle-shaped Have a single nucleus Are arranged in sheets or layers that alternately contract and relax to change the size or shape of structures	Forms the walls of hollow internal structures such as blood vessels, the gastrointestinal tract, and the bladder	Moves substances through tracts Regulates organ volume	Involuntary Contracts auto-rhythmically and is controlled by neurotransmitters and hormones

Skeletal Muscle

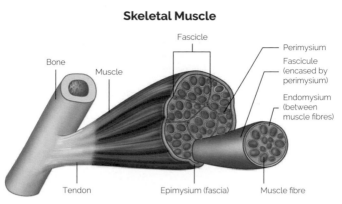

Figure 5.2: Connective tissue wrappings of a skeletal muscle

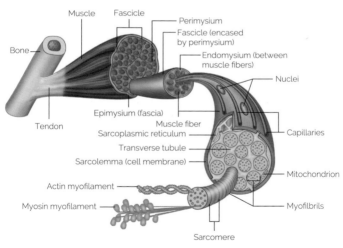

Figure 5.3: Structure of a skeletal muscle

Table 5.2: Structure of a skeletal muscle

The outer protection of a skeletal muscle: Connective tissue
Skeletal muscles work together with the skeleton to produce movement.
Holding the muscles and limbs together is a dense, irregular connective
tissue called **fascia**. Fascia separates muscles into different functional
groups and fills spaces between muscles, thus allowing free movement of
the muscles. It also supports the nerves, blood, and lymphatic vessels that
serve the muscles. Beneath the fascia lie muscles. Looking at a muscle under
a microscope, you will see hundreds of muscle fibers surrounded and held
together by connective tissue. This connective tissue surrounds, protects,
and reinforces the fibers.

Connective Tissue	Description
Epimysium	The outermost layer that encircles the whole muscle
Perimysium	Surrounds bundles of 10–100 muscle fibers. These bundles are called fascicles
Endomysium	Surrounds each individual muscle fiber within the fascicle. It contains many blood capillaries so that each muscle fiber has a good supply of blood, which brings oxygen and nutrients to the muscles and removes their waste products
Tendon	A strong cord of dense connective tissue that attaches muscles to bones, to the skin, or to other muscles
Aponeurosis	A flat, sheet-like tendon that attaches muscles to bone, to skin, or to another muscle

The inner cells of a skeletal muscle: Muscle fibers
Each muscle fiber within the fascicle is a single cell that has similar properties
to a generalized animal cell. Muscle fibers (myofibers) are long and
cylindrical in shape, multi-nucleated, and have many mitochondria.

Cellular Component	Description
Sarcolemma	Plasma membrane of a muscle fiber
Sarcoplasm	Cytoplasm of a muscle fiber
Sarcoplasmic reticulum	Stores calcium, necessary for muscular contraction
Myofibril	Long, threadlike organelle that is the contractile element of skeletal muscles. Myofibrils are made of **myofilaments** and are arranged into **sarcomeres**

Table 5.2: (continued)

Cellular Component	Description
Sarcomere	The basic functional unit of a skeletal muscle. A sarcomere contains three types of filaments, which move to overlap one another and cause a muscle to shorten: • Thick filaments contain the protein myosin • Thin filaments contain actin, tropomyosin, troponin, and myosin-binding sites • Elastic filaments contain titin (connectin) and help stabilize the thick filaments Sarcomeres are made up of two bands, which give muscles their striated appearance: • The A-band is dark and contains mainly thick filaments. In the center of the A-band is the H-zone, which contains thick filaments only. The H-zone is divided by an M-line of protein molecules, which holds the thick filaments together • The I-band is a light area containing thin filaments only Sarcomeres are separated from one another by Z-disks/lines

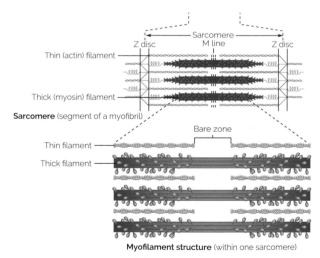

Figure 5.4: Structure of a sarcomere

Muscle contraction and relaxation

Muscles contract, or shorten, when the thick and thin filaments in the sarcomere slide past one another. This is known as the **sliding-filament mechanism**, and it occurs as follows:

- When a muscle is relaxed, the myosin-binding sites on the actin molecules are covered by a tropomyosin-troponin complex, and the myosin heads are in an energized state.
- A nerve impulse triggers the release of acetylcholine, which triggers a muscle action potential, which causes the release of calcium.
- Calcium binds with the tropomyosin-troponin complex to free up the myosin-binding sites.
- The myosin heads bind to actin with a power stroke, which draws the thin filaments inward toward the H-zone.
- The thick filaments remain in the same place and the muscle shortens (contracts).
- The myosin heads then detach from the actin and attach to another myosin-binding site further along the thin filament.
- The cycle continues as long as ATP and calcium are present.

Muscles relax when:

- Acetylcholine is broken down by an enzyme. This stops further muscle action potentials and therefore stops the release of calcium and leads to a decrease in calcium levels.

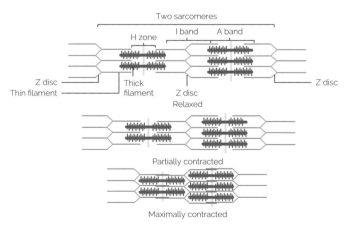

Figure 5.5: The sliding-filament mechanism

- When there is not enough calcium available to bind with the tropomyosin-troponin complex, it moves back over the myosin-binding sites and blocks the myosin heads from binding with the actin. Thus, the thin filaments slip back into their relaxed position and no more contraction takes place.

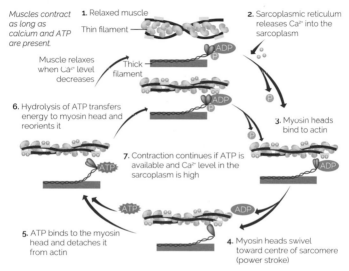

Muscles contract as long as calcium and ATP are present.

Muscle relaxes when Ca²⁺ level decreases

1. Relaxed muscle
Thin filament
Thick filament

2. Sarcoplasmic reticulum releases Ca²⁺ into the sarcoplasm

3. Myosin heads bind to actin

6. Hydrolysis of ATP transfers energy to myosin head and reorients it

7. Contraction continues if ATP is available and Ca²⁺ level in the sarcoplasm is high

5. ATP binds to the myosin head and detaches it from actin

4. Myosin heads swivel toward centre of sarcomere (power stroke)

Figure 5.6: Cycle of muscular contraction and relaxation

Tone, or tonus, is the partial contraction of a resting muscle. Even when we think our muscles are completely relaxed, a few of the fibers are still involuntarily contracted. This gives them their firmness or tension and is essential for maintaining posture.

Most physical activities include two types of contraction: **isotonic** and **isometric**.

Types of muscular contraction

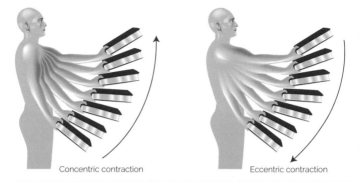

Concentric contraction   Eccentric contraction

Isotonic contractions: Isotonic contractions are regular contractions in which muscles shorten and create movement, while the tension in the muscle remains constant. They improve muscle strength and joint mobility and come in two forms:

– **Concentric contractions:** These are always toward the center and are contractions in which the muscle shortens and generates a movement that decreases the angle at a joint.
– **Eccentric contractions:** These are always away from the center and are contractions in which muscles lengthen.

Figure 5.7: Isotonic contraction

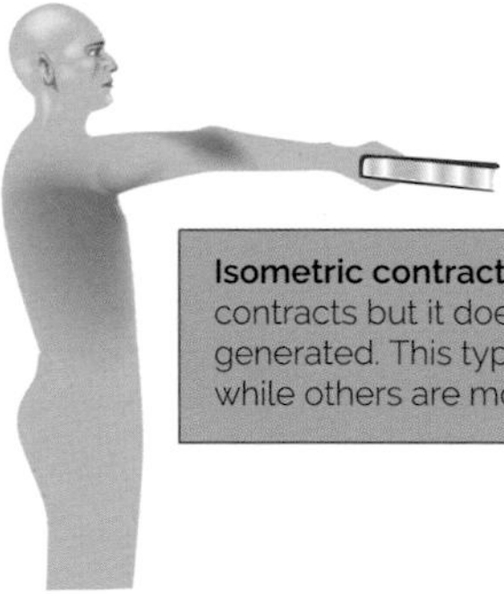

Isometric contractions: In isometric contractions, the muscle contracts but it does not shorten and no movement is generated. This type of contraction stabilizes some joints while others are moved. It also improves muscle tone.

Figure 5.8: Isometric contraction

Skeletal muscles and movement

A muscle is usually attached to two bones that form a joint—and when the muscle contracts it pulls the movable bone toward the stationary bone. All muscles have at least two attachments. The point where the muscle attaches to the stationary bone is called the "origin" and the point where the muscle attaches to the moving bone is called the "insertion." During contraction, the insertion usually moves toward the origin.

Muscles at joints are usually arranged in opposing pairs: the muscle responsible for causing a particular movement is called the **prime mover**, or **agonist**, and the muscle that opposes this movement is called the **antagonist**. Additional muscles at the joint ensure a steady movement and help the prime mover function effectively. These muscles are called **synergists**, and they are usually found alongside the prime mover. Specialized synergists—called fixators or stabilizers—stabilize the bone of the prime mover's origin so that it can act efficiently.

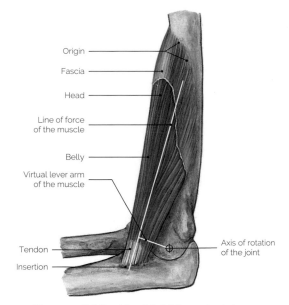

Origin

Fascia

Head

Line of force
of the muscle

Belly

Virtual lever arm
of the muscle

Tendon

Insertion

Axis of rotation
of the joint

Figure 5.9: Relationship of skeletal muscles to bones

Table 5.3: Muscle metabolism

Skeletal Muscle Metabolic Energy System	Description
Phosphagen system	Muscles use the small amount of ATP that they store in their own fibers. This is only enough energy to last for around 15 seconds of maximal muscular activity
Glycolysis	Muscles break down glucose and convert it into pyruvic acid (pyruvate) and ATP. The ATP is used by the muscles, while the pyruvic acid enters the mitochondria of the muscle fiber, where it needs oxygen to be broken down completely. If there is not enough oxygen present (anaerobic glycolysis) to completely break down the pyruvic acid, then it is converted into **lactic acid**
Aerobic respiration	Occurs if there is enough oxygen to completely break down the pyruvic acid into carbon dioxide, water, ATP, and heat. This process is called cellular respiration or biological oxidation and provides energy for activities of longer than 10 minutes, as long as there is an adequate supply of oxygen and nutrients

Table 5.4: Muscle fiber types

Features	Type I: Slow-Twitch	Type II: Fast-Twitch	
	Type I: Slow Oxidative	*Type IIA: Fast Oxidative*	*Type IIB: Fast Glycolytic*
Color	Red	Red to pink	White
Diameter of fiber	Smallest	Medium	Largest
Oxygen supply	Contains large amounts of myoglobin (therefore red), many mitochondria, and many blood capillaries	Contains large amounts of myoglobin (therefore red to pink), many mitochondria, and many blood capillaries	Has a low myoglobin content (therefore white), few mitochondria, and few blood capillaries
ATP production	Generates ATP by aerobic processes (therefore called oxidative fibers)	Generates ATP by aerobic processes (therefore called oxidative fibers)	Generates ATP by anaerobic processes (glycolysis); therefore, cannot supply the muscle continuously with ATP

Table 5.4: *(continued)*

Features	Type I: Slow-Twitch	Type II: Fast-Twitch	
	Type I: Slow Oxidative	*Type IIA: Fast Oxidative*	*Type IIB: Fast Glycolytic*
Contraction velocity	Splits ATP slowly and therefore has a slow contraction velocity	Splits ATP quickly and therefore has a fast contraction velocity	Due to its large diameter, it splits ATP very fast and therefore has a strong, rapid contraction velocity
Fatigue resistance	Very resistant to fatigue	Resistant to fatigue, but not as much as type 1	Fatigues easily
Activities	Maintaining posture and endurance activities	Walking and running	Fast movements, such as throwing a ball

Muscles of the Body

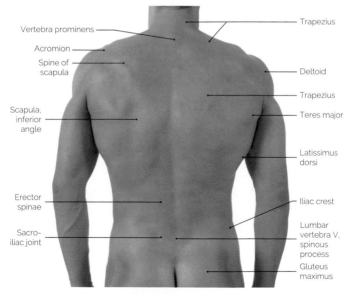

Figure 5.10: The back

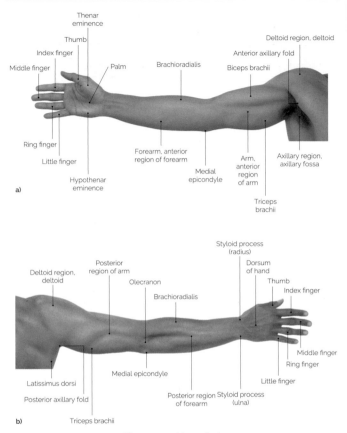

Figure 5.11: Upper limb

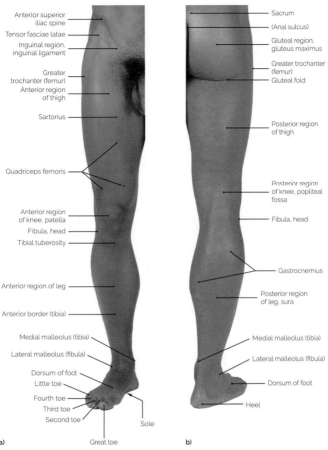

Anterior superior iliac spine

Tensor fasciae latae

Inguinal region, inguinal ligament

Greater trochanter (femur)

Anterior region of thigh

Sartorius

Quadriceps femoris

Anterior region of knee, patella

Fibula, head

Tibial tuberosity

Anterior region of leg

Anterior border (tibia)

Medial malleolus (tibia)

Lateral malleolus (fibula)

Dorsum of foot

Little toe

Fourth toe

Third toe

Second toe

Sole

Great toe

Sacrum

(Anal sulcus)

Gluteal region, gluteus maximus

Greater trochanter (femur)

Gluteal fold

Posterior region of thigh

Posterior region of knee, popliteal fossa

Fibula, head

Gastrocnemius

Posterior region of leg, sura

Medial malleolus (tibia)

Lateral malleolus (fibula)

Dorsum of foot

Heel

a)

b)

Figure 5.12: Lower limb

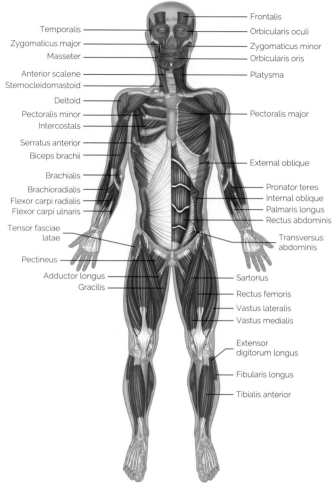

Figure 5.13: Overview of the skeletal muscles (anterior view)

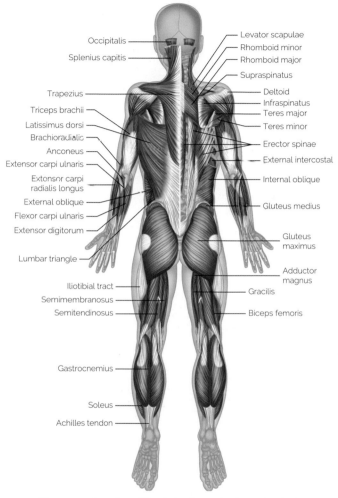

Occipitalis

Splenius capitis

Trapezius

Triceps brachii

Latissimus dorsi

Brachioradialis

Anconeus

Extensor carpi ulnaris

Extensor carpi radialis longus

External oblique

Flexor carpi ulnaris

Extensor digitorum

Lumbar triangle

Iliotibial tract

Semimembranosus

Semitendinosus

Gastrocnemius

Soleus

Achilles tendon

Levator scapulae

Rhomboid minor

Rhomboid major

Supraspinatus

Deltoid

Infraspinatus

Teres major

Teres minor

Erector spinae

External intercostal

Internal oblique

Gluteus medius

Gluteus maximus

Adductor magnus

Gracilis

Biceps femoris

Figure 5.14: Overview of the skeletal muscles (posterior view)

Muscles of the face and scalp

Procerus
Levator labii superioris
Zygomaticus minor
Zygomaticus major
Platysma (cut)
Mentalis
Sternocleidomastoid
Levator scapulae
Posterior scalene

Occipitofrontalis (frontal belly)
Temporalis
Orbicularis oculi
Masseter
Orbicularis oris
Depressor labii inferioris
Depressor anguli oris
Anterior scalene
Middle scalene
Trapezius

a)

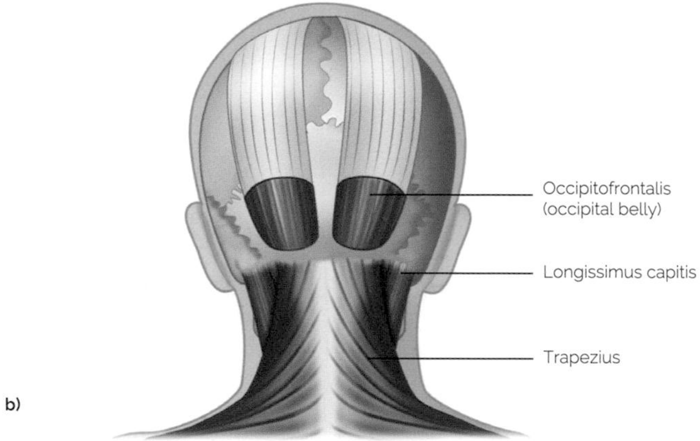

Occipitofrontalis (occipital belly)
Longissimus capitis
Trapezius

b)

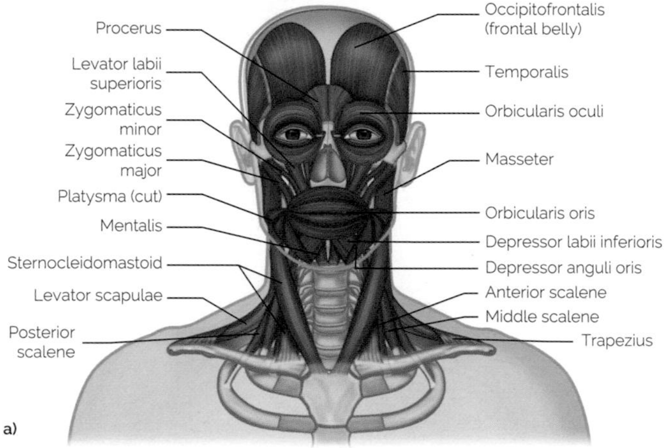

Figure 5.15 (a–b): Muscles of the face and scalp, (a) anterior view; (b) posterior view

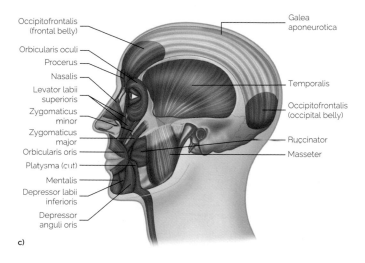

Occipitofrontalis (frontal belly)
Orbicularis oculi
Procerus
Nasalis
Levator labii superioris
Zygomaticus minor
Zygomaticus major
Orbicularis oris
Platysma (cut)
Mentalis
Depressor labii inferioris
Depressor anguli oris

Galea aponeurotica
Temporalis
Occipitofrontalis (occipital belly)
Buccinator
Masseter

c)

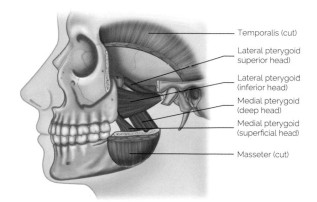

Temporalis (cut)
Lateral pterygoid superior head)
Lateral pterygoid (inferior head)
Medial pterygoid (deep head)
Medial pterygoid (superficial head)
Masseter (cut)

d)

Figure 5.15 (c–d): (c) muscles of the head and neck (posterior view); (d) muscles of mastication (lateral view)

Muscle	Origin	Insertion	Nerve	Action
Scalp				
Occipitofrontalis	*Frontal belly:* Skin of eyebrows *Occipital belly:* Lateral two-thirds of superior nuchal line of occipital bone. Mastoid process of temporal bone	Galea aponeurotica	Facial nerve (VII)	*Frontal belly:* Raises eyebrows and wrinkles skin of forehead horizontally *Occipital belly:* Pulls scalp backward
Temporoparietalis	Fascia above ear	Lateral border of galea aponeurotica	Facial nerve (VII)	Tightens scalp. Raises ears
Ear				
Superior auricular	Fascia in temporal region above ear	Superior part of ear	Facial nerve (VII)	Elevates ear
Anterior auricular	Anterior part of temporal fascia	Into helix of ear	Facial nerve (VII)	Draws ear forward and upward
Posterior auricular	Mastoid process of temporal bone	Posterior part of ear	Facial nerve (VII)	Pulls ear backward and upward
Eyelids				
Orbicularis oculi	*Orbital part:* Frontal bone. Frontal process of maxilla. Medial palpebral ligament *Palpebral part:* Medial palpebral ligament	*Orbital part:* Circular path around orbit, returning to origin *Palpebral part:* Lateral palpebral raphe	Facial nerve (VII)	*Orbital part:* Strongly closes eyelids *Palpebral part:* Gently closes eyelids
Levator palpebrae superioris	Root of orbit (lesser wing of sphenoid bone)	Skin of upper eyelid	Oculomotor nerve (III)	Raises upper eyelid
Corrugator supercilii	Medial end of superciliary arch of frontal bone	Deep surface of skin under medial half of eyebrows	Facial nerve (VII)	Draws eyebrows medially and downward
Nose				
Procerus	Fascia over nasal bone. Upper part of lateral nasal cartilage	Skin between eyebrows	Facial nerve (VII)	Produces wrinkles over bridge of nose
Nasalis	*Transverse part:* Maxilla just lateral to nose *Alar part:* Maxilla over lateral incisor	*Transverse part:* Joins muscle of opposite side across bridge of nose *Alar part:* Alar cartilage of nose	Facial nerve (VII)	*Transverse part:* Compresses nasal aperture *Alar part:* Draws cartilage downward and laterally
Depressor septi nasi	Maxilla above medial incisor	Nasal septum and ala	Facial nerve (VII)	Pulls the nose inferiorly

Muscle	Origin	Insertion	Nerve	Action
Mouth				
Depressor anguli oris	Oblique line of mandible	Skin at corner of mouth	Facial nerve (VII)	Pulls corner of mouth downward and laterally
Depressor labii inferioris	Anterior part of oblique line of mandible	Skin of lower lip	Facial nerve (VII)	Pulls lower lip downward and laterally
Mentalis	Mandible inferior to incisor teeth	Skin of chin	Facial nerve (VII)	Protrudes lower lip and pulls up skin of chin
Risorius	Fascia over masseter muscle	Skin at corner of mouth	Facial nerve (VII)	Retracts corner of mouth
Zygomaticus major	Posterior part of lateral surface of zygomatic bone	Skin at corner of mouth	Facial nerve (VII)	Pulls corner of mouth upward and laterally
Zygomaticus minor	Anterior part of lateral surface of zygomatic bone	Upper lip just medial to corner of mouth	Facial nerve (VII)	Elevates upper lip
Levator labii superioris	*Angular head:* Zygomatic bone and frontal process of maxilla *Infraorbital head:* Lower border of orbit	*Angular head:* Greater alar cartilage, upper lip, and skin of nose *Infraorbital head:* Muscles of upper lip	Facial nerve (VII)	Raises upper lip. Dilates nostril
Levator anguli oris	Canine fossa of maxilla	Skin at corner of mouth	Facial nerve (VII)	Elevates corner of mouth
Orbicularis oris	Muscle fibers surrounding opening of mouth	Skin and fascia at corner of mouth	Facial nerve (VII)	Closes lips. Protrudes lips
Buccinator	Posterior parts of maxilla and mandible; pterygomandibular raphe	Blends with orbicularis oris and into lips	Facial nerve (VII)	Presses cheek against teeth. Compresses distended cheeks
Mastication				
Masseter	Zygomatic arch and maxillary process of zygomatic bone	Lateral surface of ramus of mandible	Trigeminal nerve (V)	Elevation of mandible
Temporalis	Bone of temporal fossa. Temporal fascia	Coronoid process of mandible. Anterior margin of ramus of mandible	Trigeminal nerve (V)	Elevation and retraction of mandible

Muscle	Origin	Insertion	Nerve	Action
Mastication (continued)				
Lateral pterygoid	*Superior head:* Roof of infratemporal fossa. *Inferior head:* Lateral surface of lateral plate of pterygoid process	*Superior head:* Capsule and articular disk of temporomandibular joint. *Inferior head:* Neck of mandible	Trigeminal nerve (V)	Protrusion and side-to-side movements of mandible
Medial pterygoid	*Deep head:* Medial surface of lateral pterygoid plate of pterygoid process. Pyramidal process of palatine bone. *Superficial head:* Tuberosity of maxilla and pyramidal process of palatine bone	Medial surface of ramus and angle of mandible	Trigeminal nerve (V)	Elevation and side-to-side movement of mandible

Muscles of the neck

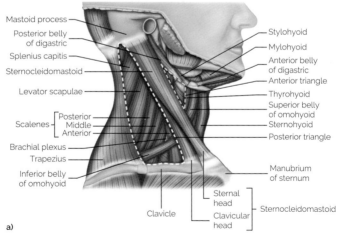

Figure 5.16a: Muscles of the neck, lateral view

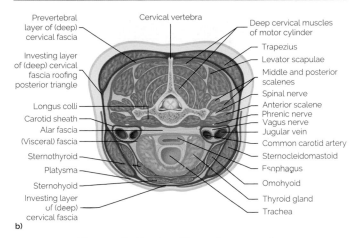

Figure 5.16b: Muscles of the neck, cross-section

Muscle	Origin	Insertion	Nerve	Action
Platysma	Subcutaneous fascia of upper quarter of chest	Subcutaneous fascia and muscles of chin and jaw. Inferior border of mandible	Facial nerve (VII)	Pulls lower lip from corner of mouth downward and laterally. Draws skin of chest upward
Anterior Triangle—Suprahyoid Muscles				
Mylohyoid	Mylohyoid line on inner surface of mandible	Median fibrous raphe and adjacent part of hyoid bone	Mylohyoid nerve from inferior alveolar branch of mandibular nerve (V₃)	Depresses mandible when hyoid is fixed. Elevates and pulls hyoid forward when mandible is fixed. Supports and elevates floor of oral cavity
Geniohyoid	Inferior mental spine on inner surface of mandible	Hyoid bone	Branch from ventral ramus of C1 carried along hypoglossal nerve (XII)	Protrudes and elevates hyoid bone. Depresses mandible if hyoid bone is fixed
Stylohyoid	Base of styloid process of temporal bone	Hyoid bone	Facial nerve (VII)	Pulls hyoid bone upward and backward, thereby elevating tongue

Muscle	Origin	Insertion	Nerve	Action
Anterior Triangle—Suprahyoid Muscles (continued)				
Digastric	*Anterior belly:* Digastric fossa on inner side of lower border of mandible.	Body of hyoid bone via a fascial sling over an intermediate tendon	*Anterior belly:* Mylohyoid nerve, from mandibular nerve (V_3).	*Anterior belly:* Raises hyoid bone. Opens mouth by lowering mandible.
	Posterior belly: Mastoid notch on medial side of mastoid process of temporal bone		*Posterior belly:* Facial nerve (VII)	*Posterior belly:* Pulls hyoid upward and back
Anterior Triangle—Infrahyoid Muscles				
Sternohyoid	Posterior aspect of sternoclavicular joint, and adjacent manubrium of sternum	Lower border of hyoid bone (medial to insertion of omohyoid)	Ventral rami of C13 through the ansa cervicalis	Depresses hyoid bone after swallowing
Sternothyroid	Posterior surface of manubrium of sternum	Oblique line on outer surface of thyroid cartilage	Ventral rami of C1–3 through the ansa cervicalis	Draws larynx downward
Thyrohyoid	Oblique line of outer surface of thyroid cartilage	Lower border of body and greater horn of hyoid bone	Fibers from ventral ramus of C1 carried along hypoglossal nerve (XII)	Raises thyroid and depresses hyoid bone
Omohyoid	*Inferior belly:* Upper border of scapula medial to the scapular notch.	*Inferior belly:* Intermediate tendon.	Ventral rami of C1 to C3 through ansa cervicalis	Depresses and fixes hyoid bone
	Superior belly: Intermediate tendon	*Superior belly:* Lower border of hyoid bone, lateral to insertion of sternohyoid		
Prevertebral and Lateral Vertebral Muscles				
Longus colli	*Superior oblique:* Transverse processes of C3–5.	*Superior oblique:* Anterior arch of atlas.	Ventral rami of cervical nerves C2–6	Flexes neck anteriorly and laterally and slight rotation to opposite side
	Inferior oblique: Anterior surface of bodies of T1, 2, maybe T3.	*Inferior oblique:* Transverse processes of C5–6.		
	Vertical: Anterior surface of bodies of T1–3 and C5–7	*Vertical:* Transverse processes of C2–4		

Muscle	Origin	Insertion	Nerve	Action
Prevertebral and Lateral Vertebral Muscles (*continued*)				
Longus capitis	Transverse processes of C3–6	Inferior surface of basilar part of occipital bone	Ventral rami of cervical nerves C1–3, (C4)	Flexes head
Rectus capitis anterior	Anterior surface of lateral mass of atlas and its transverse process	Inferior surface of basilar part of occipital bone	Branches from ventral rami of cervical nerves C1, 2	Flexes head at atlanto-occipital joint
Rectus capitis lateralis	Transverse process of atlas	Jugular process of occipital bone	Branches from ventral rami of cervical nerves C1, 2	Flexes head laterally to same side. Stabilizes atlanto-occipital joint
Posterior Triangle				
Scalenes	*Anterior:* Anterior tubercles of transverse processes of C3–6. *Middle:* Transverse processes of C2–7. *Posterior:* Posterior tubercles of transverse processes of C4–6	*Anterior:* Scalene tubercle and upper surface of 1st rib. *Middle:* Upper surface of 1st rib, behind groove for subclavian artery. *Posterior:* Upper surface of 2nd rib	*Anterior:* Ventral rami of cervical nerves C4–7. *Middle:* Ventral rami of cervical nerves C3–7. *Posterior:* Ventral rami of lower cervical nerves C5–7	*Acting on both sides:* Flex neck; raise 1st or 2nd rib during active respiratory inhalation. *Acting on one side:* Side flexes and rotates neck
Sternocleidomastoid	*Sternal head:* Upper part of anterior surface of manubrium of sternum. *Clavicular head:* Upper surface of medial third of clavicle	*Sternal head:* Lateral one-half of superior nuchal line of occipital bone. *Clavicular head:* Outer surface of mastoid process of temporal bone	Accessory nerve (XI) and branches from ventral rami of cervical nerves C2, 3 (C4)	*Bilateral contraction:* Draws head forward (protracts); raises sternum, and consequently ribs, during deep inhalation. *Unilateral contraction:* Flexes head to same side; rotates head to opposite side

Muscles of the trunk and shoulder

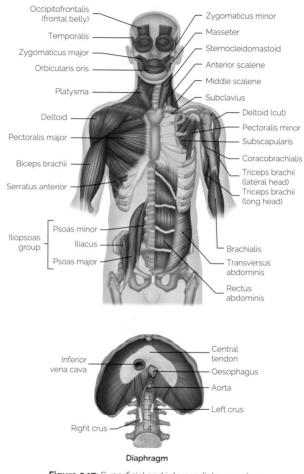

Occipitofrontalis (frontal belly)
Temporalis
Zygomaticus major
Orbicularis oris
Platysma
Deltoid
Pectoralis major
Biceps brachii
Serratus anterior
Iliopsoas group
　Psoas minor
　Iliacus
　Psoas major

Zygomaticus minor
Masseter
Sternocleidomastoid
Anterior scalene
Middle scalene
Subclavius
Deltoid (cut)
Pectoralis minor
Subscapularis
Coracobrachialis
Triceps brachii (lateral head)
Triceps brachii (long head)
Brachialis
Transversus abdominis
Rectus abdominis

Central tendon
Inferior vena cava
Oesophagus
Aorta
Left crus
Right crus

L1
L2
3
L4

Diaphragm

Figure 5.17: Superficial and intermediate muscles of the upper body (anterior view)

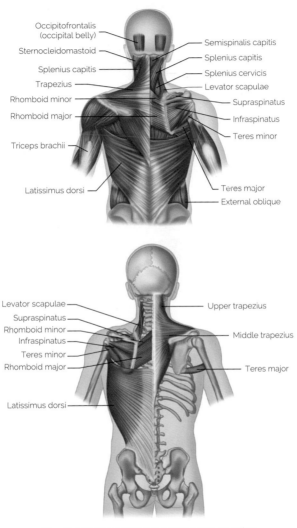

Figure 5.18: Superficial and intermediate muscles of the upper body (posterior view)

Muscle	Origin	Insertion	Nerve	Action
Muscles of the Thorax				
Intercostals	*External:* Lower border of a rib. *Internal:* Upper border of a rib and costal cartilage. *Innermost:* Superior border of each rib	*External:* Upper border of rib below. *Internal:* Lower border of rib above. *Innermost:* Inferior border of the preceding rib	The corresponding intercostal nerves	Contract to stabilize ribcage during movements of trunk. Prevent intercostal space from bulging out or sucking in during respiration. Act to fix the position of the ribs during respiration (innermost only)
Diaphragm	*Sternal portion:* Back of xiphoid process. *Costal portion:* Inner surfaces of lower 6 ribs and their costal cartilages. *Lumbar portion:* L1–3. Medial and lateral lumbocostal arches	All fibers converge and attach onto a central tendon	Phrenic nerve (ventral rami) C3–5	Forms floor of thoracic cavity. Pulls central tendon downward during inhalation
Muscles of the Anterior Abdominal Wall				
Obliques	*External:* Muscular slips from the outer surfaces of the lower 8 ribs. *Internal:* Iliac crest. Lateral two-thirds of inguinal ligament. Thoracolumbar fascia	*External:* Lateral lip of iliac crest. Aponeurosis ending in linea alba. *Internal:* Inferior borders of bottom three or four ribs. Linea alba via an abdominal aponeurosis. Pubic crest and pectineal line	*External:* Ventral rami of T5–12. *Internal:* Ventral rami of T7–12 and L1	*Both together:* Compress abdomen, helping to support abdominal viscera against pull of gravity. Contraction of one side alone side flexes trunk to that side and rotates it to the opposite side. *Internal:* Contraction of one side alone side flexes and rotates trunk
Transversus abdominis	Anterior two-thirds of iliac crest. Lateral third of inguinal ligament. Thoracolumbar fascia. Costal cartilages of lower 6 ribs	Aponeurosis ending in linea alba. Pubic crest and pectineal line	Ventral rami of T7–12 and L1	Compresses abdomen
Rectus abdominis	Pubic crest, pubic tubercle, and symphysis pubis	Anterior surface of xiphoid process. 5th to 7th costal cartilages	Ventral rami of T5–12	Flexes lumbar spine and pulls ribcage down. Stabilizes pelvis during walking
Muscles of the Posterior Abdominal Wall				
Quadratus Lumborum	Transverse process of L5 vertebra. Posterior part of iliac crest. Iliolumbar ligament.	Medial part of lower border of 12th rib. Transverse processes of L1–4.	Ventral rami of T12, L1–4.	Side flexes vertebral column. Fixes 12th rib during deep respiration. Helps extend lumbar part of vertebral column and gives it lateral stability.

Muscle	Origin	Insertion	Nerve	Action
Muscles of the Posterior Abdominal Wall (continued)				
Iliopsoas	*Psoas major:* transverse processes of L1–5. Bodies of T12–L5 and intervertebral discs between each vertebra. *Iliacus:* Superior two-thirds of iliac fossa. Anterior sacroiliac and iliolumbar ligaments. Upper lateral part of sacrum.	Lesser trochanter of femur.	*Psoas major:* ventral rami of L1–3. *Iliacus:* femoral nerve L2–4.	Main flexors of hip joint. Flex and laterally rotate thigh. Bring leg forward in walking or running.
Muscles Attaching the Upper Limb to the Trunk				
Trapezius	Medial third of superior nuchal line of occipital bone. External occipital protuberance. Ligamentum nuchae. Spinous processes and supraspinous ligaments of C7 and T1–12	Superior edge of crest of spine of scapula. Medial border of acromion. Posterior border of lateral one-third of clavicle	*Motor supply:* Accessory nerve (XI). *Sensory supply (proprioception):* Ventral rami of cervical nerves C3 and 4	Powerful elevator of the scapula; rotates the scapula during abduction of humerus above horizontal. Middle fibers retract scapula. Lower fibers depress scapula
Levator scapulae	Transverse processes of C1, 2, and posterior tubercles of transverse processes of C3, 4	Posterior surface of medial border of scapula from superior angle to root of spine of scapula	Ventral rami of C3 and C4 spinal nerves and dorsal scapular nerve (C5)	Elevates scapula. Helps retract scapula. Helps side flex neck
Rhomboids	*Minor:* Spinous processes of C7, T1. Lower part of ligamentum nuchae. *Major:* Spinous processes of T2–5 and intervening supraspinous ligaments	*Both:* Posterior surface of medial border of scapula at the root of spine of scapula. *Major only:* Continues to the inferior angle	Dorsal scapular nerve C4, 5	Elevates and retracts scapula
Serratus anterior	Lateral surfaces of upper 8 or 9 ribs and deep fascia covering the related intercostal spaces	Anterior surface of medial border of scapula	Long thoracic nerve C5–7	Rotates scapula for abduction and flexion of arm. Protracts scapula
Pectoralis minor	Outer surfaces of 3rd to 5th ribs, and fascia of the corresponding intercostal spaces	Coracoid process of scapula	Medial pectoral nerve C5, (6), 7, 8, T1	Draws tip of shoulder downward. Protracts scapula. Raises ribs during forced inspiration
Subclavius	1st rib at junction between rib and costal cartilage	Groove on inferior surface of middle one-third of clavicle	Nerve to subclavius C5, 6	Draws tip of shoulder downward. Pulls clavicle medially to stabilize sternoclavicular joint

Muscle	Origin	Insertion	Nerve	Action
Muscles Attaching the Upper Limb to the Trunk (*continued*)				
Pectoralis major	*Clavicular head:* Anterior surface of medial half of clavicle. *Sternocostal head:* Anterior surface of sternum. First 7 costal cartilages. Sternal end of 6th rib. Aponeurosis of external oblique	Lateral lip of intertubercular sulcus of humerus	Medial and lateral pectoral nerves: *Clavicular head:* C5, 6. *Sternocostal head:* C6–8, T1	Flexion, adduction, and medial rotation of arm at glenohumeral joint. *Clavicular head:* Flexion of extended arm. *Sternocostal head:* Extension of flexed arm
Latissimus dorsi	Spinous processes of lower 6 thoracic vertebrae and related interspinous ligaments; via thoracolumbar fascia to the spinous processes of lumbar vertebrae, related interspinous ligaments, and iliac crest. Lower 3 or 4 ribs	Twists to insert into the floor of intertubercular sulcus of humerus, just below the shoulder joint	Thoracodorsal nerve C6–8	Adduction, medial rotation, and extension of the arm at the glenohumeral joint. Assists in forced inspiration by raising lower ribs
Muscles of the Shoulder Joint				
Deltoid	*Anterior fibers:* Anterior border of lateral one-third of clavicle. *Middle fibers:* Lateral margin of acromion process. *Posterior fibers:* Inferior edge of crest of spine of scapula	Deltoid tuberosity of humerus	Axillary nerve C5, 6	Major abductor of the arm; anterior fibers assist in flexing the arm; posterior fibers assist in extending the arm
Supraspinatus	Medial two-thirds of supraspinous fossa of scapula and deep fascia that covers the muscle	Most superior facet on the greater tubercle of humerus	Suprascapular nerve C5, 6	Initiates abduction of arm to 15 degrees at glenohumeral joint
Infraspinatus	Medial two-thirds of infraspinous fossa of scapula and deep fascia that covers the muscle	Middle facet on posterior surface of greater tubercle of humerus	Suprascapular nerve C5, 6	Lateral rotation of arm at glenohumeral joint
Teres minor	Upper two-thirds of a strip of bone on posterior surface of scapula immediately adjacent to lateral border of scapula	Inferior facet on greater tubercle of humerus	Axillary nerve C5, 6	Lateral rotation of arm at glenohumeral joint
Subscapularis	Medial two-thirds of subscapular fossa	Lesser tubercle of humerus	Upper and lower subscapular nerves C5, 6, (7)	Medial rotation of arm at glenohumeral joint
Teres major	Oval area on lower third of posterior surface of inferior angle of scapula	Medial lip of intertubercular sulcus on anterior surface of humerus	Lower subscapular nerve C5–7	Medial rotation and extension of arm at glenohumeral joint

Muscles of the back

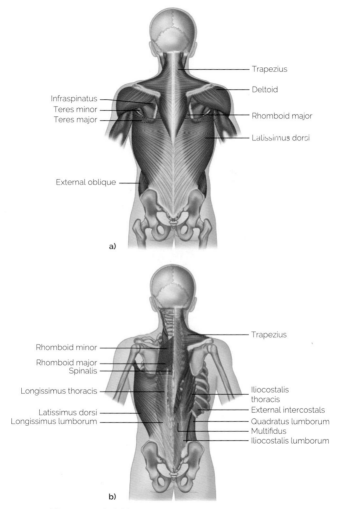

Figure 5.19 (a & b): Muscles of the trunk (posterior view)

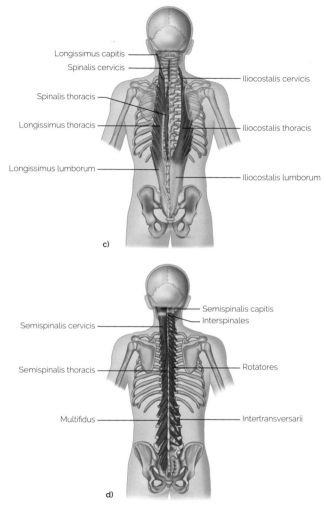

Figure 5.19 (c & d): Muscles of the trunk, (c) erector spinae muscles; (d) transversospinalis muscles

The muscles of the back can be divided into **superficial** muscles associated with movements of the shoulder; **intermediate** muscles associated with movements of the thoracic cage and respiration; and **deep** muscles associated with movements of the vertebral column. The superficial and intermediate muscles are classified as **extrinsic** muscles and are involved in moving the upper limbs and thoracic wall. The deep muscles are classified as **intrinsic** muscles and they act on the vertebral column, maintaining posture and producing movement.

The superficial extrinsic back muscles form the V-shaped musculature associated with the middle and upper back, and include the trapezius, latissimus dorsi, levator scapulae, and rhomboids. These are covered in more detail with the muscles of the trunk and shoulder on pages 112–116.

Muscle	Origin	Insertion	Nerve	Action
Spinocostal Muscles				
Serratus Posterior Superior	Lower part of ligamentum nuchae. Spinous processes of C7, T1–3. Supraspinous ligaments.	Upper borders of 2nd to 5th ribs.	Ventral rami of T2–5.	Raises upper ribs.
Serratus Posterior Inferior	Thoracolumbar fascia, at its attachment to spinous processes of T11–12 and L1–3.	Lower borders of last 4 ribs.	Ventral rami of T9–12.	May help draw lower ribs downward and backward.
Erector Spinae—Iliocostalis Portion				
Iliocostalis portion	*Lumborum:* Sacrum, spinous processes of L1–5 and T11–12 and their supraspinous ligaments. Iliac crest. *Thoracis:* Angles of lower 6 ribs. *Cervicis:* Angles of ribs 3 to 6	*Lumborum:* Angles of lower 6 or 7 ribs. *Thoracis:* Angles of upper 6 ribs and transverse process of C7. *Cervicis:* Transverse processes of C4–6	Dorsal rami of cervical, thoracic, and lumbar spinal nerves	Extends and side flexes vertebral column. Draws ribs down for forceful inhalation (thoracic only)
Erector Spinae—Longissimus Portion				
Longissimus portion	*Thoracis:* Blends with iliocostalis in lumbar region and is attached to transverse processes of lumbar vertebrae. *Cervicis:* Transverse processes of T1–5. *Capitis:* Transverse processes of T1–5. Articular processes of C4–7	*Thoracis:* Transverse processes of T1–12. Area between tubercles and angles of lower 9 or 10 ribs. *Cervicis:* Transverse processes of C2–6. *Capitis:* Posterior margin of mastoid process of temporal bone	Dorsal rami of C1–S1	Extends and side flexes vertebral column. Draws ribs down for forceful inhalation (thoracic only). Extends and rotates head (capitis only)

Muscle	Origin	Insertion	Nerve	Action
Erector Spinae—Spinalis Portion				
Spinalis portion	*Thoracic:* Spinous processes of T11–12 and L1–2. *Cervicis:* Ligamentum nuchae. Spinous process of C7. *Capitis:* Usually blends with semispinalis capitis	*Thoracis:* Spinous processes of T1–8. *Cervicis:* Spinous process of C2. *Capitis:* With semispinalis capitis	Dorsal rami of spinal nerves C2–L3	Extends vertebral column. Helps maintain correct curvature of spine in standing and sitting positions. Extends head (capitis only)
Spinotransversales Group				
Splenius capitis and splenius cervicis	*Capitis:* Lower part of ligamentum nuchae. Spinous processes of C7 and T1–4. *Cervicis:* Spinous processes of T3–6	*Capitis:* Posterior aspect of mastoid process of temporal bone. Lateral part of superior nuchal line, deep to attachment of sternocleidomastoid. *Cervicis:* Posterior tubercles of transverse processes of C1–3	*Capitis:* Dorsal rami of middle cervical nerves. *Cervicis:* Dorsal rami of lower cervical nerves	*Acting on both sides:* Extend head and neck. *Acting on one side:* Side flexes neck; rotates head to same side as contracting muscle
Transversospinales Group				
Semispinalis	*Thoracic:* Transverse processes of T6–10. *Cervicis:* Transverse processes of T1–6. *Capitis:* Transverse processes of C4–T7	*Thoracis:* Spinous processes of C6–T4. *Cervicis:* Spinous processes of C2–5. *Capitis:* Between superior and inferior nuchal lines of occipital bone	Dorsal rami of thoracic and cervical spinal nerves	Extends thoracic and cervical parts of vertebral column. Assists in rotation of thoracic and cervical vertebrae. Semispinalis capitis extends and assists in rotation of the head
Multifidus	Sacrum, origin of erector spinae, PSIS, mammillary processes of all lumbar vertebrae. Transverse processes of all thoracic vertebrae. Articular processes of lower 4 cervical vertebrae	Base of spinous processes of all vertebrae from L5 to C2	Dorsal rami of spinal nerves	Extension, side flexion, and rotation of vertebral column
Rotatores	Transverse process of each vertebra	Base of spinous process of adjoining vertebra above	Dorsal rami of spinal nerves	Rotate and assist in extension of vertebral column

Muscles of the arm and forearm

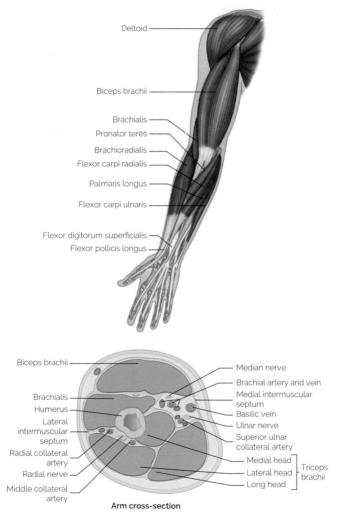

Figure 5.20 (a): Muscles of the upper limb (anterior view)

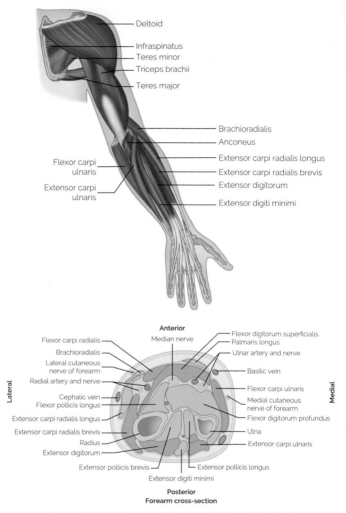

Figure 5.20 (b): Muscles of the upper limb (posterior view)

Muscle	Origin	Insertion	Nerve	Action
Muscles of the Arm—Anterior Compartment				
Biceps brachii	*Long head:* Supraglenoid tubercle of scapula. *Short head:* Tip of coracoid process	Radial tuberosity	Musculocutaneous nerve C5, 6	Powerful flexor of forearm at elbow joint. Supinates forearm
Brachialis	Anterior aspect of humerus (medial and lateral surfaces) and adjacent intermuscular septae	Tuberosity of ulna	Musculocutaneous nerve C5, 6	Flexor of forearm at elbow joint
Coracobrachialis	Tip of coracoid process	Medial aspect of humerus at mid-shaft	Musculocutaneous nerve C5–7	Flexor of arm at glenohumeral joint
Muscles of the Arm—Posterior Compartment				
Triceps brachii	*Long head:* Infraglenoid tubercle of scapula. *Medial and lateral heads:* Posterior surface of humerus	Posterior part of olecranon process of ulna	Radial nerve C6–8	Extends forearm at elbow joint
Muscles of the Anterior Compartment of the Forearm—Superficial Layer				
Flexor carpi ulnaris	*Humeral head:* Medial epicondyle. *Ulnar head:* Olecranon and posterior border of ulna	Pisiform. Hook of hamate. Base of 5th metacarpal	Ulnar nerve C7, 8, T1	Flexes and adducts wrist
Palmaris longus	Medial epicondyle of humerus	Palmar aponeurosis of hand	Median nerve C(6), 7, 8	Flexes wrist joint. Tenses palmar fascia
Flexor carpi radialis	Medial epicondyle of humerus	Bases of 2nd and 3rd metacarpals	Median nerve C6, 7	Flexes and abducts wrist joint
Pronator teres	*Humeral head:* Medial epicondyle and adjacent supra-epicondylar ridge. *Ulnar head:* Medial border of coronoid process	Mid-lateral surface of radius	Median nerve C6, 7	Pronates forearm
Muscles of the Anterior Compartment of the Forearm—Intermediate Layer				
Flexor digitorum superficialis	*Humero-ulnar head:* Medial epicondyle. Adjacent border of coronoid process. *Radial head:* Oblique line of radius	Four tendons that insert into the sides of the middle phalanges of the four fingers	Median nerve C8, T1	Flexes proximal IP joints of the index, middle, ring, and little fingers

Muscle	Origin	Insertion	Nerve	Action
Muscles of the Anterior Compartment of the Forearm—Deep Layer				
Flexor digitorum profundus	Medial and anterior surfaces of ulna. Medial half of interosseous membrane	Four tendons that attach to the palmar surfaces of the distal phalanges of the index, middle, ring, and little fingers	*Medial half:* Ulnar nerve C8, T1. *Lateral half:* Anterior interosseous branch of median nerve C8, T1	Flexes distal IP joints of the index, middle, ring, and little fingers
Flexor pollicis longus	Anterior surface of shaft of radius. Radial half of interosseous membrane	Palmar surface of base of distal phalanx of thumb	Anterior interosseous branch of median nerve C(6), 7, 8	Flexes IP joint of thumb
Pronator quadratus	Linear ridge on distal anterior surface of ulna	Distal anterior surface of radius	Anterior interosseous branch of median nerve C7, 8	Pronation
Muscles of the Posterior Compartment of the Forearm—Superficial Layer				
Brachioradialis	Proximal part of lateral supra-epicondylar ridge and adjacent intermuscular septum	Lower surface of distal end of radius	Radial nerve C5, 6	Accessory flexor of elbow joint when forearm is mid-pronated
Extensor carpi radialis longus	Distal part of lateral supra-epicondylar ridge and adjacent intermuscular septum	Dorsal surface of base of 2nd metacarpal	Radial nerve C6, 7	Extends and abducts wrist
Extensor carpi radialis brevis	Lateral epicondyle and adjacent intermuscular septum	Dorsal surface of base of 2nd and 3rd metacarpals	Radial nerve C7, 8	Extends and abducts wrist
Extensor digitorum	Lateral epicondyle and adjacent intermuscular septum and deep fascia	Four tendons that insert via extensor hoods into the dorsal aspects of the bases of the middle and distal phalanges of the index, middle, ring, and little fingers	Posterior interosseous nerve C7, 8	Extends the index, middle, ring, and little fingers
Extensor digiti minimi	Lateral epicondyle and adjacent intermuscular septum together with extensor digitorum	Extensor hood of little finger	Posterior interosseous nerve C6, 7, 8	Extends little finger

Muscle	Origin	Insertion	Nerve	Action
Muscles of the Posterior Compartment of the Forearm—Superficial Layer (continued)				
Extensor carpi ulnaris	Lateral epicondyle and posterior border of ulna	Tubercle on base of medial side of 5th metacarpal	Posterior interosseous nerve C6, 7, 8	Extends and adducts wrist
Anconeus	Lateral epicondyle	Olecranon process and proximal posterior surface of ulna	Radial nerve C6, 7, 8	Abduction of ulna in pronation. Accessory extensor of elbow joint
Muscles of the Posterior Compartment of the Forearm—Deep Layer				
Supinator	*Superficial part:* Lateral epicondyle. Radial collateral and annular ligaments. *Deep part:* Supinator crest of ulna	Lateral surface of radius superior to the anterior oblique line	Deep branch of the radial nerve C7, 8	Supination
Abductor pollicis longus	Posterior surfaces of ulna and radius. Intervening interosseous membrane	Lateral side of base of 1st metacarpal	Posterior interosseous nerve C7, 8	Abducts carpometacarpal joint of thumb; accessory extensor of thumb
Extensor pollicis brevis	Posterior surface of radius. Adjacent interosseous membrane	Base of dorsal surface of proximal phalanx of thumb	Posterior interosseous nerve C7, 8	Extends MCP joint of thumb
Extensor pollicis longus	Posterior surface of ulna. Adjacent interosseous membrane	Dorsal surface of base of distal phalanx of thumb	Posterior interosseous nerve C7, 8	Extends IP joint of thumb
Extensor indicis	Posterior surface of ulna. Adjacent interosseous membrane	Extensor hood of index finger	Posterior interosseous nerve C7, 8	Extends index finger

Muscles of the hand

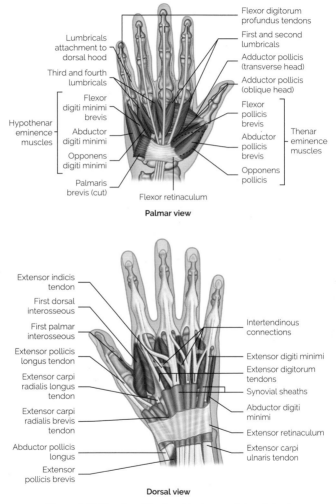

Flexor digitorum profundus tendons

First and second lumbricals

Adductor pollicis (transverse head)

Adductor pollicis (oblique head)

Flexor pollicis brevis

Abductor pollicis brevis

Opponens pollicis

Thenar eminence muscles

Lumbricals attachment to dorsal hood

Third and fourth lumbricals

Flexor digiti minimi brevis

Abductor digiti minimi

Opponens digiti minimi

Palmaris brevis (cut)

Hypothenar eminence muscles

Flexor retinaculum

Palmar view

Extensor indicis tendon

First dorsal interosseous

First palmar interosseous

Extensor pollicis longus tendon

Extensor carpi radialis longus tendon

Extensor carpi radialis brevis tendon

Abductor pollicis longus

Extensor pollicis brevis

Intertendinous connections

Extensor digiti minimi

Extensor digitorum tendons

Synovial sheaths

Abductor digiti minimi

Extensor retinaculum

Extensor carpi ulnaris tendon

Dorsal view

Figure 5.21: Muscles of the hand (palmar and dorsal view)

Muscle	Origin	Insertion	Nerve	Action
Muscles of the Hand				
Palmaris brevis	Palmar aponeurosis. Flexor retinaculum	Skin on ulnar border of hand	Superficial branch of ulnar nerve C(7), 8, T1	Improves grip
Dorsal interossei	Adjacent sides of metacarpals	Extensor hood and base of proximal phalanges of index, middle, and ring fingers	Deep branch of ulnar nerve C8, T1	Abduction of index, middle, and ring fingers at MCP joints
Palmar interossei	Sides of metacarpals	Extensor hoods of the thumb, index, ring, and little fingers and proximal phalanx of thumb	Deep branch of ulnar nerve C8, T1	Adduction of the thumb, index, ring, and little fingers at MCP joints
Adductor pollicis	*Transverse head:* Palmar surface of 3rd metacarpal *Oblique head:* Capitate and bases of 2nd and 3rd metacarpals	Base of proximal phalanx of thumb and extensor hood of thumb	Deep branch of ulnar nerve C8, T1	Adducts thumb
Lumbricals	Tendons of flexor digitorum profundus	Extensor hoods of index, ring, middle, and little fingers	*Lateral lumbricals:* Digital branches of median nerve. *Medial lumbricals:* Deep branch of ulnar nerve	Extend IP joints and simultaneously flex MCP joints
Muscles of the Hand—Hypothenar Eminence				
Abductor digiti minimi	Pisiform, pisohamate ligament, and tendon of flexor carpi ulnaris	Proximal phalanx of little finger	Deep branch of ulnar nerve C(7), 8, T1	Abducts little finger at MCP joint
Opponens digiti minimi	Hook of hamate. Flexor retinaculum	Entire length of medial border of 5th metacarpal	Deep branch of ulnar nerve C(7), 8, T1	Laterally rotates 5th metacarpal
Flexor digiti minimi brevis	Hook of hamate. Flexor retinaculum	Proximal phalanx of little finger	Deep branch of ulnar nerve C(7), 8, T1	Flexes little finger at MCP joint
Muscles of the Hand—Thenar Eminence				
Abductor pollicis brevis	Tubercles of trapezium and scaphoid and adjacent flexor retinaculum	Proximal phalanx and extensor hood of thumb	Recurrent branch of median nerve C8, T1	Abducts thumb at MCP joint
Opponens pollicis	Flexor retinaculum. Tubercle of trapezium	Entire length of radial border of 1st metacarpal	Recurrent branch of median nerve C8, T1	Medially rotates thumb
Flexor pollicis brevis	Flexor retinaculum. Tubercle of trapezium	Proximal phalanx of thumb	Recurrent branch of median nerve C8, T1	Flexes thumb at MCP joint

Muscles of the hip and thigh

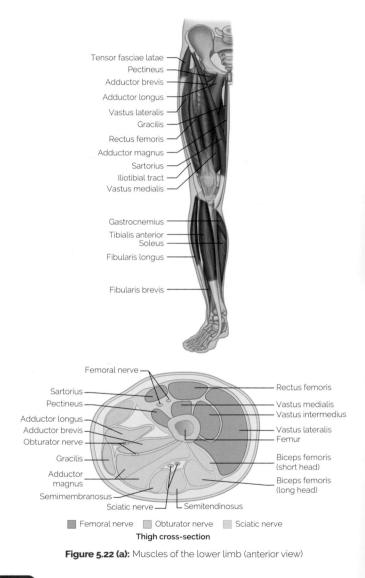

Figure 5.22 (a): Muscles of the lower limb (anterior view)

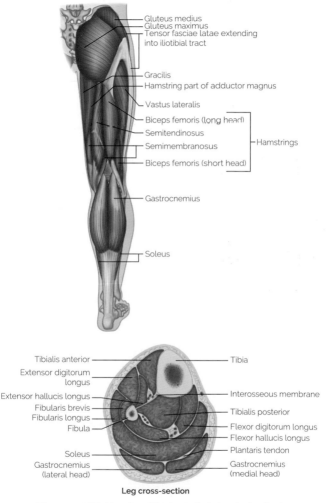

Gluteus medius
Gluteus maximus
Tensor fasciae latae extending into iliotibial tract
Gracilis
Hamstring part of adductor magnus
Vastus lateralis
Biceps femoris (long head)
Semitendinosus
Semimembranosus
Biceps femoris (short head)
Hamstrings
Gastrocnemius
Soleus

Tibialis anterior
Extensor digitorum longus
Extensor hallucis longus
Fibularis brevis
Fibularis longus
Fibula
Soleus
Gastrocnemius (lateral head)
Tibia
Interosseous membrane
Tibialis posterior
Flexor digitorum longus
Flexor hallucis longus
Plantaris tendon
Gastrocnemius (medial head)

Leg cross-section

Figure 5.22 (b): Muscles of the lower limb (posterior view)

Muscle	Origin	Insertion	Nerve	Action
Muscles of the Gluteal Region				
Gluteus maximus	Outer surface of ilium and posterior surface of sacrum and coccyx (over sacroiliac joint)	Posterior aspect of ITT. Gluteal tuberosity of proximal femur	Inferior gluteal nerve L5, S1, 2.	Powerful extensor of flexed femur at hip joint. Lateral stabilizer of hip and knee joints. Laterally rotates and abducts thigh
Tensor fasciae latae	Lateral aspect of crest of ilium between ASIS and tubercle of the crest	ITT	Superior gluteal nerve L4, 5, S1	Stabilizes the knee in extension
Gluteus medius	External surface of ilium between anterior and posterior gluteal lines	Oblique ridge on lateral surface of greater trochanter	Superior gluteal nerve L4, 5, S1	Abducts femur at hip joint. Medially rotates thigh
Gluteus minimus	External surface of ilium between anterior and inferior gluteal lines	Anterolateral border of greater trochanter	Superior gluteal nerve L4, 5, S1	Abducts, medially rotates, and may assist in flexion of hip joint
Piriformis	Anterior surface of sacrum between anterior sacral foramina	Medial side of superior border of greater trochanter	Branches from sacral nerves S1, 2	Laterally rotates extended femur at hip joint. Abducts flexed femur at hip joint
Deep lateral hip rotators	*Obturator internus:* Inner surface of ischium, pubis and ilium. *Gemellus superior:* External surface of ischial spine. *Gemellus inferior:* Upper aspect of ischial tuberosity. *Quadratus femoris:* Lateral edge of ischium just anterior to ischial tuberosity	Greater trochanter of femur (except quadratus femoris, which inserts just behind and below the others)	*Obturator internus and gemellus superior:* Nerve to obturator internus, L5, S1. *Gemellus inferior and quadratus femoris:* Nerve to quadratus femoris, L5, S1, (2)	Laterally rotates hip joint. Abducts flexed femur at hip joint. Helps hold head of femur in acetabulum
Muscles of the Anterior Compartment of the Thigh				
Sartorius	ASIS	Medial surface of tibia just inferomedial to tibial tuberosity	Femoral nerve L2, 3, (4)	Flexes the thigh at the hip joint. Flexes the leg at the knee joint
Quadriceps femoris	*Rectus femoris:* Straight head: AIIS; reflected head: groove above acetabulum (on ilium). *Vasti group:* Upper half of shaft of femur	Patella, then via patellar ligament into the tibial tuberosity	Femoral nerve L2, 3, 4	*Rectus femoris:* Flexes the thigh at the hip joint and extends leg at the knee joint. *Vasti group:* Extend leg at the knee joint

Muscle	Origin	Insertion	Nerve	Action
Muscles of the Medial Compartment of the Thigh				
Gracilis	A line on the external surfaces of the pubis, the inferior pubic ramus, and ramus of the ischium	Medial surface of proximal shaft of tibia	Obturator nerve L2, 3	Adducts thigh at hip joint. Flexes leg at knee joint
Pectineus	Pecten pubis and adjacent bone of pelvis	Oblique line, from base of lesser trochanter to linea aspera of femur	Femoral nerve L2, 3	Adducts and flexes thigh at hip joint
Obturator externus	External surface of obturator membrane and adjacent bone	Trochanteric fossa	Posterior division of obturator nerve L3, 4	Laterally rotates thigh at hip joint
Adductors	Anterior part of the pubic bone (ramus). Adductor magnus also takes its origin from the ischial tuberosity	Entire length of femur, along linea aspera and medial supracondylar line to adductor tubercle on medial epicondyle of femur	*Magnus:* Obturator nerve L2, 3, 4. Sciatic nerve L2, 3, 4. *Brevis:* Obturator nerve L2, 3. *Longus:* Obturator nerve L2, 3, 4	Adduct and medially rotate thigh at hip joint
Muscles of the Posterior Compartment of the Thigh				
Hamstrings	Ischial tuberosity. *Biceps femoris (short head only):* Lateral lip of linea aspera	*Semimembranosus:* Groove and adjacent bone on medial and posterior surface of medial tibial condyle. *Semitendinosus:* Medial surface of proximal tibia. *Biceps femoris:* Head of fibula	Sciatic nerve L5, S1, 2	Flexes leg at knee joint. Semimembranosus and semitendinosus extend thigh at hip joint, medially rotate thigh at hip joint and leg at knee joint. Biceps femoris extends and laterally rotates thigh at hip joint and laterally rotates leg at knee joint

Muscles of the leg

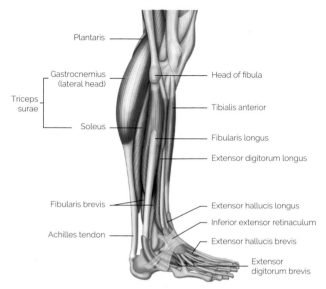

Figure 5.23: Muscles of the leg (lateral view)

Muscle	Origin	Insertion	Nerve	Action
Muscles of the Anterior Compartment of the Leg				
Tibialis anterior	Lateral surface of tibia and adjacent interosseous membrane	Medial and inferior surfaces of medial cuneiform and adjacent surfaces on base of 1st metatarsal	Deep fibular nerve L4, 5	Dorsiflexes foot at ankle joint. Inverts foot
Extensor digitorum longus	Proximal one-half of medial surface of fibula and related surface of lateral tibial condyle	Along dorsal surface of the 4 lateral toes. Each tendon divides, to attach to bases of middle and distal phalanges	Deep fibular nerve L5, S1	Extends lateral four toes and dorsiflexes foot
Extensor hallucis longus	Middle one-half of medial surface of fibula and adjacent interosseous membrane	Base of distal phalanx of great toe	Deep fibular nerve L5, S1	Extends great toe. Dorsiflexes foot
Fibularis tertius	Distal part of medial surface of fibula	Dorsomedial surface of base of 5th metatarsal	Deep fibular nerve L5, S1	Dorsiflexes and everts foot

Muscle	Origin	Insertion	Nerve	Action
Muscles of the Posterior Compartment of the Leg—Superficial Layer				
Gastrocnemius	*Medial head:* Posterior surface of distal femur just superior to medial condyle. *Lateral head:* Upper posterolateral surface of lateral femoral condyle	Posterior surface of calcaneus via the Achilles tendon	Tibial nerve S1, 2	Plantar flexes foot. Flexes knee
Soleus	Posterior aspect of fibular head and adjacent surfaces of neck and proximal shaft. Soleal line and medial border of tibia. Tendinous arch between tibial and fibular attachments	Posterior surface of calcaneus via the Achilles tendon	Tibial nerve S1, 2	Plantar flexes foot
Plantaris	Lower part of lateral supracondylar line of femur and oblique popliteal ligament of knee joint	Posterior surface of calcaneus via the Achilles tendon	Tibial nerve S1, 2	Plantar flexes foot. Flexes knee
Muscles of the Posterior Compartment of the Leg—Intermediate Layer				
Flexor digitorum longus	Medial side of posterior surface of tibia, below soleal line	Plantar surfaces of bases of distal phalanges of lateral 4 toes	Tibial nerve S2, 3	Flexes lateral four toes
Flexor hallucis longus	Lower two-thirds of posterior surface of fibula and adjacent interosseous membrane	Plantar surface of base of distal phalanx of great toe	Tibial nerve S2, 3	Flexes great toe and is important in the final propulsive thrust of foot during walking
Muscles of the Posterior Compartment of the Leg—Deep Layer				
Tibialis posterior	Posterior surfaces of interosseous membrane and adjacent regions of tibia and fibula	Mainly to tuberosity of navicular and adjacent region of medial cuneiform	Tibial nerve L4, 5	Inverts and plantar flexes foot
Popliteus	Lateral femoral condyle	Posterior surface of proximal tibia	Tibial nerve L4, 5, S1	Stabilizes and unlocks the knee joint
Muscles of the Lateral Compartment of the Leg				
Fibularis longus	Upper two-thirds of lateral surface of fibular head, fibula, and occasionally lateral tibial condyle	Lateral side of distal end of medial cuneiform. Base of 1st metatarsal	Superficial fibular nerve L5, S1, 2	Everts and plantar flexes foot
Fibularis brevis	Lower two-thirds of lateral surface of shaft of fibula	Lateral tubercle at base of 5th metatarsal	Superficial fibular nerve L5, S1, 2	Everts foot

Muscles of the foot

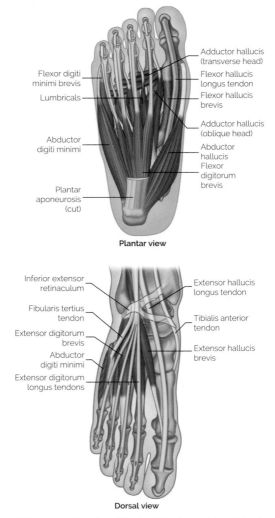

Plantar view

Flexor digiti minimi brevis
Lumbricals
Abductor digiti minimi
Plantar aponeurosis (cut)

Adductor hallucis (transverse head)
Flexor hallucis longus tendon
Flexor hallucis brevis
Adductor hallucis (oblique head)
Abductor hallucis
Flexor digitorum brevis

Dorsal view

Inferior extensor retinaculum
Fibularis tertius tendon
Extensor digitorum brevis
Abductor digiti minimi
Extensor digitorum longus tendons

Extensor hallucis longus tendon
Tibialis anterior tendon
Extensor hallucis brevis

Figure 5.24: Muscles of the foot (plantar and dorsal view)

Muscle	Origin	Insertion	Nerve	Action
Muscles of the Sole of the Foot—First Layer				
Abductor hallucis	Medial process of calcaneal tuberosity	Medial side of base of proximal phalanx of great toe	Medial plantar nerve from tibial nerve S1–3	Abducts and flexes great toe at MTP joint
Flexor digitorum brevis	Medial process of calcaneal tuberosity and plantar aponeurosis	Sides of plantar surfaces of middle phalanges of lateral 4 toes	Medial plantar nerve from tibial nerve S1–3	Flexes lateral four toes at proximal IP joint
Abductor digiti minimi	Lateral and medial processes of calcaneal tuberosity, and band of connective tissue connecting calcaneus with base of 5th metatarsal	Lateral side of base of proximal phalanx of little toe	Lateral plantar nerve from tibial nerve S1–3	Abducts 5th toe at MTP joint
Muscles of the Sole of the Foot—Second Layer				
Quadratus Plantae	Medial surface of calcaneus and lateral process of calcaneal tuberosity	Lateral border of tendon of flexor digitorum longus in proximal sole of foot	Lateral plantar nerve from tibial nerve S1–3	Flexes distal phalanges of 2nd to 5th toes
Lumbricals	*1st lumbrical:* Medial side of tendon of flexor digitorum longus associated with 2nd toe. *2nd to 4th lumbricals:* Adjacent tendons of flexor digitorum longus	Medial free margins of extensor hoods of 2nd to 5th toes	*1st lumbrical:* Medial plantar nerve from tibial nerve. *Lateral three lumbricals:* Lateral plantar nerve from tibial nerve S2, 3	Flex MTP joint and extend IP joints
Muscles of the Sole of the Foot—Third Layer				
Flexor hallucis brevis	Medial part of plantar surface of cuboid, and adjacent part of lateral cuneiform. Tendon of tibialis posterior	Lateral and medial sides of base of proximal phalanx of great toe	Medial plantar nerve from tibial nerve S1, 2	Flexes MTP joint of great toe
Adductor hallucis	*Transverse head:* Ligaments associated with MTP joints of lateral 3 toes. *Oblique head:* Bases of 2nd to 4th metatarsals; sheath covering fibularis longus tendon	Lateral side of base of proximal phalanx of great toe	Lateral plantar nerve from tibial nerve S2, 3	Adducts great toe at MTP joint
Flexor digiti minimi brevis	Base of 5th metatarsal and sheath of fibularis longus tendon	Lateral side of base of proximal phalanx of little toe	Lateral plantar nerve from tibial nerve S2, 3	Flexes little toe at MTP joint

Muscle	Origin	Insertion	Nerve	Action
Muscles of the Sole of the Foot—Fourth Layer				
Dorsal interossei	Sides of adjacent metatarsals	Extensor hoods and bases of proximal phalanges of 2nd to 4th toes	Lateral plantar nerve from tibial nerve; 1st and 2nd dorsal interossei also innervated by deep fibular nerve S2, 3	Abduct 2nd to 4th toes at MTP joints. Resist extension of MTP joints and flexion of IP joints
Plantar interossei	Bases and medial sides of 3rd–5th metatarsals	Extensor hoods and bases of proximal phalanges of 3rd–5th toes	Lateral plantar nerve from tibial nerve S2, 3	Adduct third to fifth toes at MTP joints. Resist extension of MTP joints and flexion of IP joints
Muscles of the Dorsal Aspect of the Foot				
Extensor digitorum brevis	Superolateral surface of calcaneus	Lateral sides of tendons of extensor digitorum longus of 2nd to 4th toes	Deep fibular nerve S1, 2	Extends 2nd to 4th toes
Extensor hallucis brevis	Superolateral surface of calcaneus	Base of proximal phalanx of great toe	Deep fibular nerve S1, 2	Extends MTP joint of great toe

Principal skeletal muscles

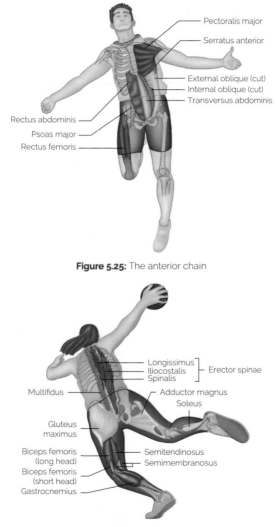

Figure 5.25: The anterior chain

Figure 5.26: The posterior chain

Theory in practice

All your body's movements are created by tiny filaments sliding to overlap one another. As they overlap, they cause a muscle to shorten, and this creates movement—of the bones and joints, internal tracts, or even the heart. It is interesting to note that muscles can only shorten. This means that every movement you make is a pull, not a push. Even when you are pushing something like a shopping trolley, your muscles are actually still pulling on bones, not pushing.

Even when we think our muscles are completely relaxed, a few of the muscle fibers are still involuntarily contracted. This constant contraction gives muscles their firmness, or tension, and is called "muscle tone," or "tonus." Tone is essential for maintaining posture and keeping our bodies in an upright position. When a person dies the muscles of the body stiffen. This is known as "rigor mortis" and it occurs because at death calcium leaks out of the sarcoplasmic reticulum and binds with troponin, triggering muscular contraction. However, because the person is dead, there is no ATP available to activate the release of the myosin heads from the actin and so the muscles are in a permanent state of contraction.

Skeletal-muscle cells do not have much potential to divide but, if injured, they can be replaced by new cells derived from dormant stem cells called "satellite cells." However, if there is more damage to the muscle than the satellite cells can cope with, fibrosis occurs. This is the replacement of muscle fibers by scar tissue, which is a fibrous connective tissue that does not allow much movement, and which can restrict movement at joints. Adhesions are bands of scar tissue that join two surfaces of the body that are usually separate.

CHAPTER 6

The Nervous System

The nervous system is made up of millions of nerve cells that all communicate with one another to control the body and maintain homeostasis.

A stimulus is something that provokes a response, be it a change in the temperature of the air or a pin prick to a finger. **Sensory receptors** pick up stimuli from both inside and outside the body. All the information gathered by these sensory receptors is called **sensory input**.

Once the nervous system has new sensory input, it **analyzes**, **processes** and *interprets* this information in the brain and spinal cord. It is also able to store the information and make decisions regarding it. Having picked up a change in the environment and decided what to do about it, the final function of the nervous system is the ability to **act**, or **respond** to a stimulus, by glandular secretions or muscular contractions. This is called **motor output**.

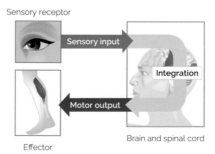

Figure 6.1: Functions of the nervous system

We do not always control our skeletal muscles voluntarily. Sometimes they contract involuntarily through what is called a **reflex arc**.

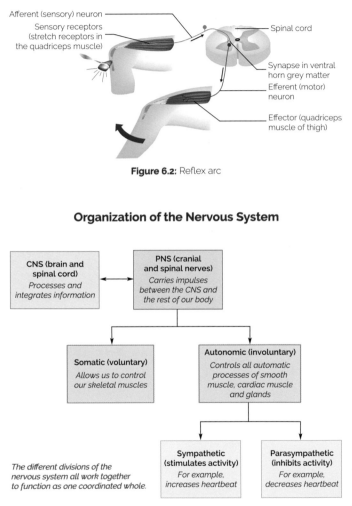

Figure 6.2: Reflex arc

Organization of the Nervous System

Figure 6.3: Divisions of the nervous system

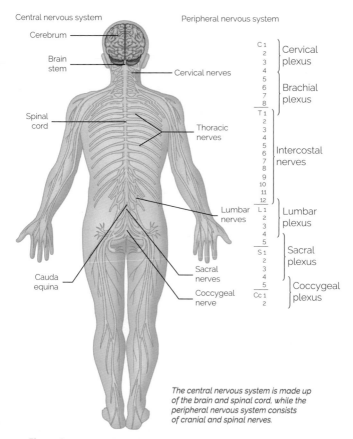

Central nervous system

Cerebrum

Brain stem

Spinal cord

Cauda equina

Peripheral nervous system

Cervical nerves

Thoracic nerves

Lumbar nerves

Sacral nerves

Coccygeal nerve

C 1
2
3
4
5
6
7
8
} Cervical plexus

} Brachial plexus

T 1
2
3
4
5
6
7
8
9
10
11
12
} Intercostal nerves

L 1
2
3
4
5
} Lumbar plexus

S 1
2
3
4
5
} Sacral plexus

Cc 1
2
} Coccygeal plexus

The central nervous system is made up of the brain and spinal cord, while the peripheral nervous system consists of cranial and spinal nerves.

Figure 6.4: An overview of the central and peripheral nervous systems

Table 6.1: Effects of the sympathetic and parasympathetic nervous systems

Sympathetic Stimulation	Structure	Parasympathetic
Pupil dilated	Iris muscle	Pupil constricted
Vasoconstriction	Blood vessels in head	No effect
Secretion inhibited	Salivary glands	Secretion increased
Rate and force of contraction increased	Heart	Rate and force of contraction decreased
Vasodilation	Coronary arteries	Vasoconstriction
Bronchodilation	Trachea and bronchi	Bronchoconstriction
Peristalsis reduced, sphincters closed	Stomach	Secretion of gastric juice increased
Glycogen to glucose conversion increased	Liver	Blood vessels dilated, secretion of bile increased

Table 6.1: (continued)

Sympathetic Stimulation	Structure	Parasympathetic
Epinephrine (adrenaline) and norepinephrine (noradrenaline) secreted into blood	Adrenal medulla	No effect
Peristalsis reduced, sphincters closed	Large and small intestines	Secretions and peristalsis increased, sphincter relaxed
Smooth muscle wall relaxed, sphincter closed	Bladder	Smooth muscle wall contracted, sphincter relaxed

Nervous Tissue

The nervous system contains only two types of cells: **neuroglia** and **neurons**. Neuroglia—or glia—are smaller and more numerous than neurons and are the "glue," or supporting cells, of nervous tissue. They insulate, nurture, and protect neurons and maintain homeostasis of the fluid surrounding neurons.

There are six different types of neuroglia, but they all have two things in common: they cannot transmit nerve impulses and they can divide by mitosis. Of the six different types of neuroglia, four of them are present only in the CNS. These are astrocytes, oligodendrocytes, microglia, and ependymal cells. The remaining two are present in the PNS and are Schwann cells and satellite cells.

Neurons are the cells responsible for the sensory, integrative and motor functions of the nervous system. They can differ in size and shape, but they all share one essential characteristic: they transmit impulses or electrical signals to, from, or within the brain.

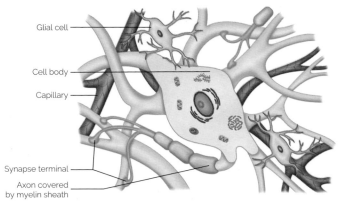

Glial cell

Cell body

Capillary

Synapse terminal

Axon covered
by myelin sheath

Figure 6.5: Neuroglia and neurons

Table 6.2: Classification of neurons

Neuron	Information Carried	Direction	
		From:	To:
Sensory/afferent	Sensory nerve impulse	Skin, sense organs, muscle, joints, viscera	CNS
Motor/efferent	Motor nerve impulse	CNS	Muscles or glands (called effectors)
Association/ interneurons	These are not specifically sensory or motor neurons; rather, they connect sensory and motor neurons in neural pathways		

Structure of a Motor Neuron

All neurons have three parts: a **cell body**, which is the metabolic center of the neuron; **dendrites**, which receive information; and an **axon**, which transmits information. Some axons are covered in a sheath which protects and insulates the neuron and speeds up the conduction of nerve impulses. This sheath is called a **myelin sheath**.

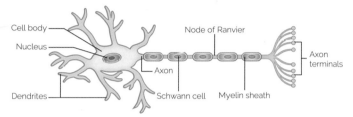

Figure 6.6: Structure of a motor neuron

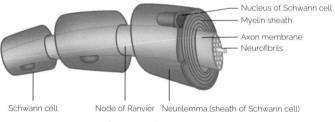

Figure 6.7: A myelinated fiber

Table 6.3: Structure of a motor neuron

Name	Description	
Cell body (soma)	The metabolic center of the neuron. It is similar to a generalized animal cell but does not include centrioles, and therefore division via mitosis is not possible. The cell body also has organelles specific to neurons only: • **Nissl bodies**—made of rough endoplasmic reticulum and are the site of protein synthesis • **Neurofibrils**—form the cytoskeleton of the cell to maintain its shape	
Dendrites	The receiving or input portion of the cell. They are branching processes that project from the cell body and are short and unmyelinated. Neurons can have many dendrites projecting from their cell body	
Axon	The transmitting portion of a cell. It transmits nerve impulses away from the cell body and toward another neuron, muscle fiber, or gland cell. An axon looks like a long tail and the end of it divides into many fine processes called axon terminals, which contain membrane-enclosed sacs called **synaptic vesicles**. Synaptic vesicles store and release **neurotransmitters** into the **synaptic cleft**. Some axons are **myelinated**: they are covered in a **myelin sheath**, which protects and insulates the axon and speeds up the conduction of the nerve impulse. Myelin is produced by Schwann cells in the PNS and oligodendrocytes in the CNS	

Transmission of a Nerve Impulse

Nerve impulses are transmitted **electrochemically** either across the plasma membrane of an unmyelinated axon or across the nodes of Ranvier of a myelinated axon.

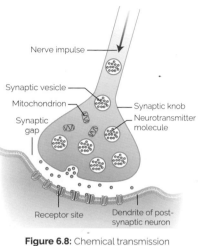

Figure 6.8: Chemical transmission across the synapse

A nerve impulse is generated and propagated as follows:

- An inactive plasma membrane has a resting membrane potential that is **polarized**. This means that there is an electrical voltage difference across the membrane, with the external voltage being positive and the internal one being negative. The main external ions are sodium, while the main internal ones are potassium.
- When the dendrites of the neuron are **stimulated**, ion channels in a small segment of the plasma membrane open and allow the movement of sodium ions into the cell. This causes the inside of the cell to become positive and the outside negative. This is called "**depolarization**."
- Depolarization causes the membrane potential to be reversed and this initiates an **action potential (impulse)**. When one segment of the membrane becomes depolarized, it causes the segment next to it to be depolarized, and so a wave of depolarization is propagated down the length of the plasma membrane. This is how the action potential travels to the end of the neuron.
- When the action potential reaches the end of the neuron, it causes vesicles containing neurotransmitters to open up and release a neurotransmitter into the synaptic cleft. The neurotransmitter diffuses across the synaptic cleft and binds to the receptors of the next neuron, muscle, or gland, where it now acts as a stimulus.

Nerves

A nerve consists of a bundle of nerve fibers surrounded by connective tissue.

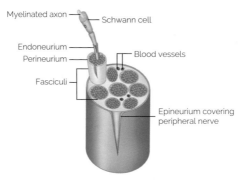

Figure 6.9: Structure of a nerve

Table 6.4: Classification of nerves

Type	Description
Sensory or afferent nerve	Contains sensory fibers that carry sensory impulses toward the CNS
Motor or efferent nerve	Contains motor fibers that carry motor impulses away from the CNS
Mixed nerve	Contains both sensory and motor fibers. All spinal nerves are mixed nerves

Brain

The brain lies in the cranial cavity and is composed of four regions: the **brain stem** which is continuous with the spinal cord and consists of the medulla oblongata, pons Varolii and midbrain; the **cerebellum** which is found at the back of the head behind the brain stem; the **diencephalon** which lies above the brain stem and includes the epithalamus, thalamus and hypothalamus; and the **cerebrum** which covers the diencephalon, is composed of an outer cerebral cortex and inner white matter and basal ganglia, and is divided into the right and left cerebral hemispheres.

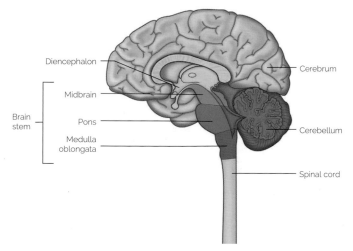

Figure 6.10: The human brain

Table 6.5: Regions of the brain

Region	Description and Functions
Brain stem	A continuation of the spinal cord and connects the spinal cord to the diencephalon. Its main function is to relay motor and sensory impulses between the spinal cord and the other parts of the brain
Medulla oblongata (often referred to as the medulla)	Found at the top of the spinal cord Approximately 1¼ in. (3 cm) long

Table 6.5: (*continued*)

Region	Description and Functions
The medulla oblongata contains:	

All sensory and motor white matter **tracts that connect the brain with the spinal cord;** most of these tracts cross over from the left to the right and vice versa, so the left side of the brain controls the muscles of the right side of the body and vice versa

A cardiovascular center that regulates the **heartbeat** and **diameter of blood vessels**

A medullary rhythmicity area that regulates **breathing**

Centers for the coordination of **swallowing, vomiting, coughing, sneezing,** and **hiccupping**

Nuclei of origin for **cranial nerves VIII–XII;** cranial nerves will be discussed later in this chapter

Neurons that function in **precise voluntary movements, posture,** and **balance**

Region	Description and Functions
Pons varolii (pons)	Lies above the medulla oblongata and in front of the cerebellum Approximately 1 in. (2.5 cm) long

Pons means "bridge" and the pons acts as a bridge between the spinal cord and brain as well as between different parts of the brain itself; the pons also contains:

Nuclei of origin for **cranial nerves V–VIII** (note: cranial nerve VIII has its origins in both the medulla oblongata and the pons)

Areas that function with the medullary rhythmicity area to help control **respiration**

Region	Description and Functions
Midbrain (mesencephalon)	Lies between the pons and the diencephalon Approximately 1 in. (2.5 cm) long

Table 6.5: (continued)

Region	Description and Functions
The midbrain contains white matter tracts and gray matter nuclei that function as:	

Reflex centers for **movements of the eyes, head, and neck** in response to **visual and other stimuli**

Reflex centers for **movements of the head and trunk** in response to **auditory stimuli**

Areas that control **subconscious muscle activities**

Areas that function with the basal ganglia and cerebellum to help coordinate **muscular movements**

Nuclei of origin for **cranial nerves III and IV**

Region	Description and Functions
Reticular formation	This is a mass of gray matter that extends the entire length of the brain stem

The reticular formation functions in:

The **motor control of visceral organs and in muscle tone**

Consciousness and awakening from sleep

Region	Description and Functions
Cerebellum	The second largest portion of the brain. It is what gives us the ability to perform complex movements like somersaulting, dancing, or throwing a ball. Found at the back of the head, behind the medulla oblongata and pons and beneath the cerebrum

Its main functions include:

Coordinating and smoothing complex sequences of skeletal muscular contraction

- It receives input from **proprioceptors** in muscles, tendons, and joints as well as from receptors for equilibrium and visual receptors in the eye
- It takes this information and compares the intended movement that has been programmed in the cerebrum with what is actually happening and so is able to smooth and coordinate movements

Regulating posture and balance

Table 6.5: (continued)

Region	Description and Functions
Diencephalon (interbrain)	Lies above the brain stem, where it is enclosed by the cerebral hemispheres
Epithalamus	Forms the roof of the third ventricle and is found above the thalamus
The epithalamus contains the **pineal gland** and the **choroid plexus**	
Thalamus	The word *thalamus* means "inner chamber"
	It is approximately 1¼ in. (3 cm) long and makes up about 80% of the diencephalon
	It encloses the third ventricle of the brain

Functions of the thalamus include:

It is the main **relay station** for **sensory impulses to the cerebral cortex**; sensory impulses include those of **hearing, vision, taste, touch, pressure, vibration, heat, cold**, and **pain**

It contains an area for the **crude appreciation of sensations** such as pain, temperature, and pressure before they are relayed to the cerebral cortex, where the sensations are refined

It contains nuclei that play a role in **voluntary motor actions** and **arousal**

It contains nuclei for certain **emotions** and **memory**, as well as for **cognition**, which is the ability to acquire knowledge

Hypothalamus	Found below the thalamus

Helps regulate **homeostasis**; contains receptors that monitor **osmotic pressure, hormone concentrations**, and **blood temperature**; and is connected to the endocrine system. Its functions include:

Regulating the ANS: By controlling the contraction of smooth and cardiac muscle and the secretions of many glands, the hypothalamus regulates visceral activities such as heart rate and the movement of food through the gastrointestinal tract

Table 6.5: (continued)

Region	Description and Functions
	Controlling the pituitary gland: • It releases hormones that control the secretions of the pituitary gland • It also synthesizes two hormones that are transported to, and stored in, the posterior pituitary gland until they are released; these are **oxytocin** and **antidiuretic hormone**

Controlling the pituitary gland:

• It releases hormones that control the secretions of the pituitary gland
• It also synthesizes two hormones that are transported to, and stored in, the posterior pituitary gland until they are released; these are **oxytocin** and **antidiuretic hormone**

Regulating emotional behavior: Working together with the limbic system, the hypothalamus functions in regulating emotional behavior such as rage, aggression, pain, pleasure, and sexual arousal

Regulating eating and drinking: The hypothalamus controls sensations of hunger, fullness, and thirst

Controlling body temperature

Regulating sleeping patterns

Region	Description and Functions
Cerebrum 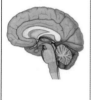	Gives us the ability to read, write, speak, remember, create, and imagine The cerebrum's outer, most superficial layer is made up of ridges and grooves that look like deep wrinkles. This layer is called the **cerebral cortex** (the word *cortex* means "rind") and it consists of **gray matter**. Beneath the gray matter of the cerebral cortex is an inner layer of **white matter**, the **limbic system**, and the **basal ganglia**

The ridges of the cortex are called **gyri** or **convolutions**, its shallow grooves are **sulci**, and its deep grooves are **fissures**. A deep fissure called the **longitudinal fissure** separates the cerebrum into two halves: the right and left hemisphere. Although they appear to be separate, these hemispheres are still connected internally by the **corpus callosum**, which is a large bundle of white-matter transverse fibers. Each hemisphere is subdivided into four lobes named after the bones that cover them: the **frontal**, **parietal**, **temporal**, and **occipital** lobes

Table 6.5: (continued)

Region	Description and Functions
Cerebral cortex	Approximately the size and shape of two closed fists held together
	Consists of gray matter ridged and grooved and separated into two hemispheres

The cerebral cortex is large and has many areas related to different functions—a simplistic view of it is that it receives almost all the sensory impulses of the body and interprets them into meaningful patterns of recognition and awareness; it has:

Sensory areas that receive and process information; these are mainly found in the posterior half of the hemispheres:

- **General sensory area** for receiving impulses related to touch, proprioception, pain, and temperature; the body can be mapped onto this sensory area, and the exact part of the body where the sensation is originating can be pinpointed
- **Visual area** for receiving information regarding characteristics of visual stimuli such as shape, color, and movement
- **Auditory area** for receiving information regarding characteristics of sound such as rhythm and pitch
- **Gustatory area** for taste
- **Olfactory area** for smell

Motor areas that output information; these are mainly found in the anterior portion of each hemisphere:

- **Area for the voluntary contraction of specific muscles or muscle groups**
- **Speech area** that translates spoken or written words into thoughts and then into speech

Association areas that consist of both sensory and motor areas; these are found mainly on the lateral surfaces of the cerebral cortex:

- The **somatosensory association area** receives, integrates, and interprets physical sensations; it also stores memories of past experiences for comparison with new information, enabling you to interpret complicated sensations, such as being able to determine the shape and texture of an object without actually looking at it
- The **visual association area** enables you to relate present sensations to past experiences and therefore be able to recognize objects

Table 6.5: (continued)

Region	Description and Functions
	• The **auditory association area** enables you to relate present sensations to past experiences and therefore be able to recognize a sound and determine whether it is speech, music, or simply a noise • The **gnostic area** integrates information and enables you to develop thoughts from a variety of sensory inputs such as smell, taste, etc. • The **premotor area** enables you to perform complicated learned motor activities such as writing • The **frontal eye field area** controls scanning movements of the eye, enabling you to perform activities such as scanning a paragraph of writing for a specific word • The **language areas** coordinate the muscles associated with speech and breathing so that you can speak
White matter	Lies beneath the cerebral cortex
Consists of axons that transmit nerve impulses around the cortex and between the brain and spinal cord	
Limbic system	Found on the inner border of the cerebrum, the floor of the diencephalon, and encircling the brain stem
Often called the "emotional brain" because it controls the emotional and involuntary aspects of behavior; it is the area associated with **pain**, **pleasure**, **anger**, **rage**, **fear**, **sorrow**, **sexual feelings**, and **affection**; it also functions in **memory**	
Basal ganglia	Groups of nuclei found in the cerebral hemispheres; they are interconnected by many nerve fibers
Receive information from, and provide output to, the cerebral cortex, thalamus, and hypothalamus. Control large **automatic movements of the skeletal muscles** and also help regulate muscle tone	

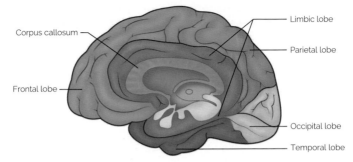

Figure 6.11: The lobes of the cerebrum (medial view)

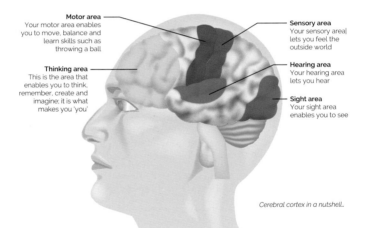

Figure 6.12: Cerebral cortex

Protection of the brain

The brain is protected in a number of ways: the hard bones of the cranium form a nearly impenetrable wall against the external environment; three layers of connective tissue—called **meninges**—cover the brain and provide further protection; and the brain itself floats in cerebrospinal fluid, which acts as a shock absorber and provides a barrier against any substances trying to enter the brain from the internal environment.

Cerebrospinal fluid (CSF) is formed from blood plasma at the choroid plexuses, which contain specialized cells called **ependymal cells**. These cells filter the blood plasma and remove any potentially harmful substances from it. CSF flows through and fills cavities in the brain called **ventricles**.

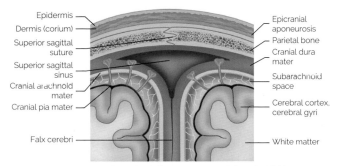

Epidermis
Dermis (corium)
Superior sagittal suture
Superior sagittal sinus
Cranial arachnoid mater
Cranial pia mater
Falx cerebri

Epicranial aponeurosis
Parietal bone
Cranial dura mater
Subarachnoid space
Cerebral cortex, cerebral gyri
White matter

The brain is protected by the scalp, skull, meninges and cerebrospinal fluid.

Figure 6.13: Protection of the brain

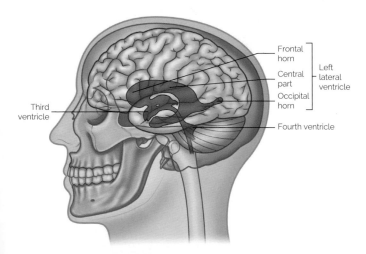

Frontal horn
Central part
Occipital horn
Left lateral ventricle
Third ventricle
Fourth ventricle

Figure 6.14: Ventricles of the brain

Cranial Nerves

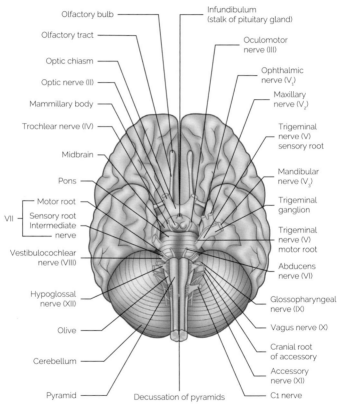

Olfactory bulb

Olfactory tract

Optic chiasm

Optic nerve (II)

Mammillary body

Trochlear nerve (IV)

Midbrain

Pons

Motor root
VII — Sensory root
Intermediate nerve

Vestibulocochlear nerve (VIII)

Hypoglossal nerve (XII)

Olive

Cerebellum

Pyramid

Infundibulum (stalk of pituitary gland)

Oculomotor nerve (III)

Ophthalmic nerve (V_1)

Maxillary nerve (V_2)

Trigeminal nerve (V) sensory root

Mandibular nerve (V_3)

Trigeminal ganglion

Trigeminal nerve (V) motor root

Abducens nerve (VI)

Glossopharyngeal nerve (IX)

Vagus nerve (X)

Cranial root of accessory

Accessory nerve (XI)

C1 nerve

Decussation of pyramids

Figure 6.15: An overview of the brain and cranial nerves

Table 6.6: Cranial nerves

Number	Name	Function
I	Olfactory	Smell
II	Optic	Vision
III	Oculomotor	Movement of eyelid and eyeball; control of lens shape and pupil size; carries autonomic nerves
IV	Trochlear	Movement of eyeball
V	Trigeminal *Three branches: Ophthalmic, maxillary, and mandibular*	**Motor function (mandibular branch only):** Chewing **Sensory function:** Sensations of touch, pain, and temperature from the skin of the face and the mucosa of the nose and mouth, and sensations supplied by proprioceptors in the muscles of mastication
VI	Abducens	Movement of eyeball
VII	Facial *Five branches: Temporal, zygomatic, buccal, mandibular, and cervical*	Facial expression; carries autonomic fibers for secretion of saliva and tears
VIII	Vestibulocochlear *Two branches: Vestibular and cochlear*	**Vestibular branch:** Balance **Cochlear branch:** Hearing
IX	Glossopharyngeal	Sensory to oropharynx; carries autonomic supply for secretion of saliva and sensory from carotid body and sinus
X	Vagus	**Motor function:** Secretion of digestive fluids and contraction of smooth muscle of organs of the thoracic and abdominal cavities **Sensory function:** Sensory input from the organs of the thoracic and abdominal cavities
XI	Accessory *Two portions: Cranial and spinal*	**Cranial portion:** Joins the vagus nerve to supply motor to pharyngeal muscles **Spinal portion:** Movement of head, sternocleidomastoid and trapezius muscles
XII	Hypoglossal	Movement of tongue when talking and swallowing

Spinal Cord

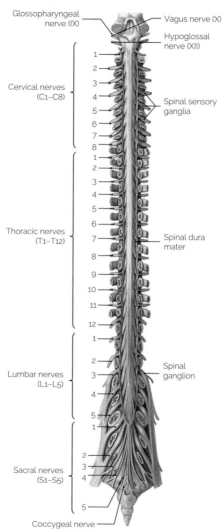

Figure 6.16: The spinal cord

The spinal cord is continuous with the brain stem and ends just above the second lumbar vertebra (L2). It transports **nerve impulses** from the periphery of the body to the brain and from the brain to the periphery. It also receives and integrates information and produces **reflex actions**, which are predictable, automatic responses to specific changes in the environment.

The spinal cord is protected by the spinal meninges, which are continuous with the cranial meninges, and the spinal cord consists of gray H-shaped (butterfly-shaped) matter surrounded by white matter.

The **gray matter** receives and integrates information and consists of cell bodies and unmyelinated axons and dendrites of association and motor neurons. Gray matter also contains nuclei where some nerve impulses are processed. The **white matter** contains tracts of myelinated fibers that transport impulses between the brain and the periphery.

Ascending tracts consist of sensory axons, which conduct nerve impulses to the brain, and descending tracts consist of motor axons, which conduct nerve impulses to the body.

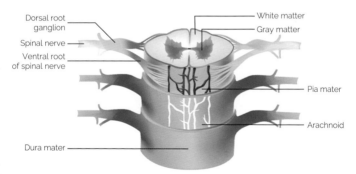

Dorsal root ganglion

Spinal nerve

Ventral root of spinal nerve

White matter

Gray matter

Pia mater

Arachnoid

Dura mater

Figure 6.17: Cross-section through the spinal cord

Spinal nerves

There are 31 pairs of nerves that originate in the spinal cord and emerge through the intervertebral foramina. These nerves form part of the PNS and connect the CNS to receptors in the muscles and glands. Each nerve has two points of attachment to the spinal cord. These attachments are called the **posterior (dorsal) root, which** consists of sensory axons, and the anterior **(ventral) root, which** consists of motor axons.

The spinal nerves are named and numbered according to where in the vertebral column they emerge. The first spinal nerve starts between the occipital and the first cervical vertebra. The lower lumbar, sacral, and coccygeal roots are not in line with their origins or corresponding vertebrae because they descend in the form of the cauda equina.

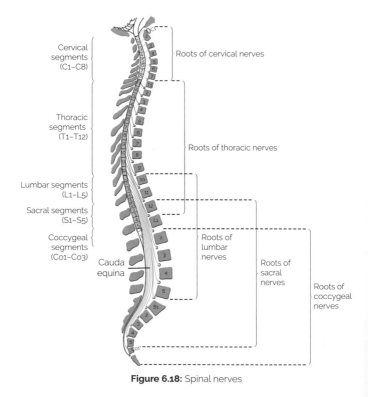

Figure 6.18: Spinal nerves

Branches of some of the spinal nerves form networks of nerves on both sides of the body. These networks are called **plexuses**, and all the nerves emerging from a particular plexus will innervate specific structures. Nerves T2–T12 do not form a plexus. These are the **intercostal nerves** that serve the muscles between the ribs, and the skin and the muscles of the anterior and lateral trunk.

Table 6.7: Spinal nerves

Plexus and Body Areas Served	Important Nerves
Cervical Skin and muscles of the head, neck, and top of the shoulders	• **Phrenic nerve** supplies motor fibers to the diaphragm
Brachial Shoulder and upper limb	• **Axillary nerve** supplies the deltoid and teres minor muscles • **Musculocutaneous nerve** supplies the flexors of the arm • **Radial nerve** supplies the muscles on the posterior aspect of the arm and forearm • **Median nerve** supplies the muscles on the anterior aspect of the forearm and some muscles of the hand • **Ulnar nerve** supplies some of the muscles of the forearm and most of the muscles of the hand
Lumbar Abdominal wall, external genitals, and part of the lower limb	• **Femoral nerve** supplies the flexor muscles of the thigh and extensor muscles of the leg, as well as the skin over parts of the thigh, leg, and foot • **Obturator nerve** supplies the adductor muscles of the leg as well as the skin over the medial aspect of the thigh
Sacral Buttocks, perineum, and lower limbs	• **Sciatic nerve** descends through the thigh and splits into the tibial and common fibular nerves

Special Sense Organs

The eye

How do we see?

If light passes through substances of different density, its speed will change and its rays will bend. This is called "**refraction**," and it takes place as light passes through the differing densities of the cornea, aqueous humor, lens, and vitreous humor. By the time the light rays reach the retina they have bent to such an extent that the image is reversed from left to right and turned upside down. The photoreceptors on the retina then convert the light into nerve impulses, which are transported by the optic nerve to the visual cortex of the brain. Here the impulses are integrated and interpreted into visual images.

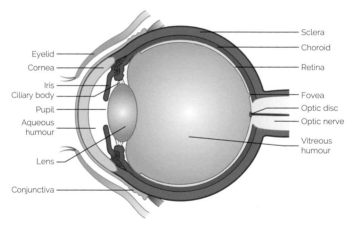

Figure 6.19: The eye

Table 6.8: Structures of the eyeball

Exterior of the Eyeball	
Structure	**Details**
Fibrous tunic • The outermost covering of the eyeball • The front of the fibrous tunic is the cornea and the back of it is the sclera 	**Cornea:** The cornea is an avascular, transparent coat that covers the iris; it is curved and helps focus light **Sclera:** The sclera is the white of the eye and is made up of dense connective tissue; it protects the eyeball and gives it shape and rigidity
Vascular tunic • The middle layer of the wall of the eyeball • The front of the vascular tunic is the colored iris, which is surrounded by the ciliary body, which then becomes the choroid 	**Iris:** This is suspended between the cornea and lens and is the colored portion of the eye; it is shaped like a flattened doughnut and the hole in the center of it is the **pupil**, which regulates the amount of light entering the eyeball; the iris consists of circular and radial smooth muscle fibers **Ciliary body:** This contains processes that secrete aqueous humor and muscles that alter the shape of the lens for near or far vision **Choroid:** This lines most of the internal surface of the sclera, is highly vascularized, and provides nutrients to the retina; it also absorbs scattered light
Nervous tunic (retina) • The innermost layer of the wall of the eyeball • It lines the posterior ¾ of the eyeball • It consists of a non-visual pigmented portion and a neural portion	**Pigmented portion:** This contains melanin and absorbs stray light rays; it therefore prevents any scattering or reflection of light within the eyeball and ensures a clear, sharp image **Neural portion:** This contains three layers of neurons that process visual input.

Table 6.8: (continued)

Exterior of the Eyeball	
Structure	**Details**
• The pigmented portion lies between the choroid and the neural portion	In these layers are the **photoreceptors**, which are specialized cells that convert light into nerve impulses; **rods**, which respond to different shades of gray only; and **cones**, which respond to color. From the photoreceptors, the information passes through bipolar cells into ganglion cells. The axons of the ganglion cells then extend into the **optic nerve**

Interior of the Eye	
Structure	**Details**
Lens: Behind the colored iris and the dark pupil is the transparent, avascular lens	The lens is responsible for fine-tuning of focusing and it is made up of protein
Anterior cavity: In front of the lens is the anterior cavity, filled with aqueous humor	**Aqueous humor** is a watery fluid that nourishes the lens and cornea and helps produce intraocular pressure; the aqueous humor is constantly replaced
Posterior cavity (vitreous chamber): This is the largest cavity in the eyeball and lies behind the lens; it contains the vitreous body (humor)	**Vitreous body (humor)** is a jelly-like substance that helps produce intraocular pressure, which helps maintain the shape of the eyeball and keeps the retina pressed firmly against the choroid; unlike the aqueous humor, the vitreous humor is not constantly replaced

The ear

How do we hear?

An object vibrates and creates alternating compressions and decompressions of air molecules. These are sound waves. Sound waves travel through the air from their source to the auricle of our ears. The auricle then directs the sound waves into the auditory canal, where they are channelled toward the eardrum. Once they hit the eardrum they cause it to vibrate, and this sets up a chain reaction of vibrations that eventually causes changes in the pressure of the endolymph in the cochlear duct. As this pressure rises and falls it moves the basilar membrane, which vibrates and causes the hair cells of the organ of Corti to move against the tectorial membrane.

The hair cells are mechanoreceptors, and they convert this mechanical stimulus into an electrical one. The hair cells synapse with neurons of cranial nerve VIII, the vestibulocochlear nerve, which sends the impulses to the medulla oblongata. The fibers then extend into the thalamus, from where the auditory signals are projected to the primary auditory area of the temporal lobe of the cerebral cortex.

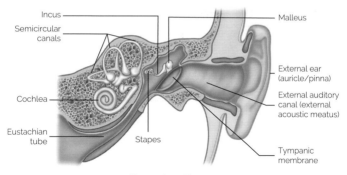

Figure 6.20: The ear

Table 6.9: Structure of the ear

Structure of the Outer Ear	
The outer—or external—ear collects and channels sound waves inward. It is composed of the auricle, external auditory canal, and eardrum. It also contains **ceruminous glands**—which secrete **cerumen** (earwax)—which, along with hairs that are also found in the outer ear, cleans and protects the ear and prevents dust and foreign objects from entering it	

Auricle (pinna)	The part of the ear we see, composed of elastic cartilage covered by skin; the upper rim is called the **helix**, while the lower lobe is called the **lobule**
External auditory canal (meatus)	A curved tube approximately 1 in. (2.5 cm) long, it carries sound waves from the auricle to the eardrum
Eardrum (tympanic membrane)	A very thin, semi-transparent membrane between the auditory canal and the middle ear; when sound waves hit it, it vibrates, passing them on to the middle ear

Structure of the Middle Ear
The middle ear is a small, air-filled cavity found between the outer ear and the inner ear. It is partitioned from the outer ear by the eardrum and from the inner ear by a bony partition containing two windows: the **oval window** and the **round window**. It contains the three auditory ossicles and is connected to the throat via the Eustachian tube

Auditory ossicles	These are three tiny bones extending across the middle ear, and each one is named after its shape: • **Hammer (malleus)** • **Anvil (incus)** • **Stirrup (stapes)** The handle of the hammer is attached to the inner surface of the eardrum, and when the eardrum vibrates it causes the hammer to move; the hammer hits the anvil, which in turn hits the stirrup, which is attached to the oval window, a membrane-covered opening that transmits the sound wave into the inner ear
Eustachian tube (auditory tube)	The Eustachian tube connects the middle ear with the upper portion of the throat (nasopharynx), and it functions in equalizing the middle-ear-cavity pressure with the external atmospheric pressure

Table 6.9: (*continued*)

Structure of the Inner Ear (Labyrinth)	
The inner, or internal, ear is sometimes also called the labyrinth. It consists of a bony labyrinth filled with a fluid called **perilymph** enclosing a membranous labyrinth filled with a fluid called **endolymph**. This labyrinth is divided into the **vestibule**, three **semicircular canals**, and the **cochlea**. Inside the cochlea lies the actual organ of hearing itself: the **organ of Corti**.	
Vestibule	This is the central portion of the bony labyrinth, and contains receptors for equilibrium
Semicircular canals	These project from the vestibule and also contain receptors for equilibrium
Cochlea	• The cochlea is a bony, spiral canal that resembles a snail's shell • It is divided into three channels, one of which is the **cochlear duct**, which is separated from another channel by the **basilar membrane** • Resting on the basilar membrane is the organ of hearing, which is called the organ of Corti, or the **spiral organ**
Organ of Corti	• The organ of Corti is a coiled sheet of epithelial cells containing thousands of hair cells • Extensions from these hair cells extend into the endolymph of the cochlear duct and are the receptors for auditory sensations • Over the hair cells is a very thin, flexible membrane called the **tectorial membrane**

The mouth

How do we taste?

The mouth houses **gustatory receptors** in **taste buds**, which are located mainly on the tongue, the back of the roof of the mouth, and in the pharynx and larynx. A taste bud is an oval body that contains gustatory receptor cells. Each of these cells has a single hairlike projection, which projects from the receptor cell and through a small opening in the taste bud called the **taste pore**.

Once inside the mouth, food is dissolved in saliva, and the hairs of gustatory receptors dip into this saliva and are stimulated by the taste chemical within it. Thus, gustatory receptors are chemoreceptors. The taste receptors then send messages to the areas of the brain responsible for taste, appetite, and saliva production. On the tongue, taste buds are found in elevations called **papillae**. These elevations give the tongue its rough surface.

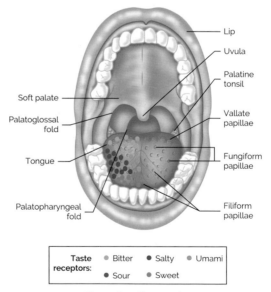

Figure 6.21: The mouth

The nose

How do we smell?

At the top of the nasal cavity is a small patch about the size of a postage stamp that contains olfactory receptors. Olfactory receptors are neurons that have many hairlike projections extending from their dendrites.

In the connective tissue that supports the olfactory epithelium are **olfactory glands (Bowman's glands)** that produce mucus. When you breathe in, airborne particles dissolve in this mucus and so come into contact with the hairlike cilia of the olfactory receptors. The olfactory cells are chemoreceptors and convert the chemical stimulus found dissolved in the mucus into nerve impulses. These impulses are then sent to the frontal lobe via the limbic system of the brain.

The sensory experience of eating involves both our senses of smell and taste, and what we know as "flavor" is actually the perception of smell and taste together. The senses of smell and taste are also unique in that their impulses travel via the limbic system to the cerebral cortex. This means that they are closely connected to our emotions and memories.

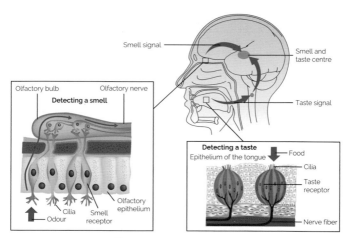

Figure 6.22: Smell and taste

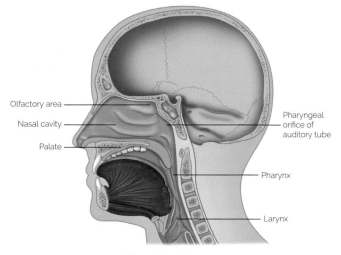

Olfactory area

Nasal cavity

Palate

Pharyngeal orifice of auditory tube

Pharynx

Larynx

Figure 6.23: The nose

Theory in practice

Neurons can vary from a millimeter to a meter in length and can conduct impulses at speeds ranging from one to more than a hundred meters per second. This is because the speed of propagation of an impulse is not related to the strength of the stimulus and is instead proportional to the diameter of the fiber, the thickness of the myelin sheath, and the temperature. The larger the diameter, the more myelin present, and the warmer the temperature, the faster the impulse will be propagated. Propagation of an impulse can be slowed down or even partially blocked by low temperatures. This is why applying ice to a painful area reduces the sensation of pain.

The fibers in our body that let us know about potential danger—for example, those that relay impulses associated with touch, pressure, and heat—have the largest diameter and therefore the quickest rate of conduction, while the fibers related to the autonomic system and therefore the involuntary, automatic functioning of the body have the slowest rate of conduction.

CHAPTER 7

The Endocrine System

Endocrine glands secrete hormones into the extracellular space around their cells. Hormones then diffuse into blood capillaries and are transported by the blood to target cells, where they bind with receptors and act like switches that turn on chemical and metabolic processes within the cell. These processes coordinate body functions such as growth, development, reproduction, metabolism, and homeostasis.

Hormone secretion is controlled by signals from the nervous system, chemical changes in the blood, and other hormones, and once hormones have been released into the body, their levels are controlled by a negative feedback mechanism. Together, the nervous and endocrine systems control all the processes that take place in the body, and to a certain extent they control one another.

Table 7.1: Comparison of the nervous and endocrine systems

Characteristic	Nervous System	Endocrine System
Messenger	Nerve impulse	Hormone
Transportation	Nerve axons	Blood
Cells affected	Mainly muscles, glands, and other neurons	All types of cells
Action	Muscular contractions and glandular secretions	All types of changes in metabolic activities, growth, development, and reproduction
Time to act	Very quick (milliseconds)	Slower (from seconds to days)
Duration of effects	Brief	Longer

Endocrine Glands and Their Hormones

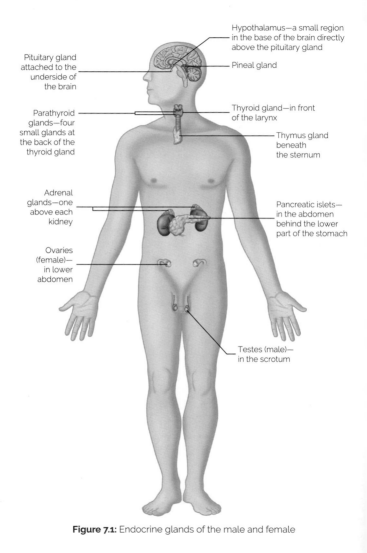

Hypothalamus—a small region in the base of the brain directly above the pituitary gland

Pineal gland

Pituitary gland attached to the underside of the brain

Parathyroid glands—four small glands at the back of the thyroid gland

Thyroid gland—in front of the larynx

Thymus gland beneath the sternum

Adrenal glands—one above each kidney

Pancreatic islets—in the abdomen behind the lower part of the stomach

Ovaries (female)—in lower abdomen

Testes (male)—in the scrotum

Figure 7.1: Endocrine glands of the male and female

Table 7.2: Endocrine glands and their hormones

Hormone	Target Tissues and Actions	Disorders and Diseases
Hypothalamus		

Not always considered an endocrine gland as it is part of the nervous system. It is included here briefly because of its vital connection to the pituitary gland, and you can read more about it in chapter 6. The hypothalamus:

- Releases a number of hormones that control the secretions of the pituitary gland
- Synthesizes two hormones that are then transported to, and stored in, the posterior pituitary gland. These hormones are oxytocin and antidiuretic hormone

Pituitary Gland		

Located in the hypophyseal fossa of the sphenoid bone and found behind the nose and between the eyes. Attached to the hypothalamus by a stalk and comprises an anterior and a posterior portion

Anterior pituitary gland
All the hormones released by the anterior pituitary gland, except for human growth hormone, regulate other endocrine glands

Hormone	Target Tissues and Actions	Disorders and Diseases
Human growth hormone (hGH), or somatotropin	Bone, muscle, cartilage, and other tissue Stimulates growth and regulates metabolism	**Hyposecretion:** Pituitary dwarfism **Hypersecretion:** Gigantism, acromegaly
Thyroid-stimulating hormone (TSH), or thyrotropin	Thyroid gland Controls the thyroid gland	**Hyposecretion:** Myxedema **Hypersecretion:** Graves' disease (NB: These diseases are more commonly caused by a problem with the thyroid gland itself)

Table 7.2: (continued)

Hormone	Target Tissues and Actions	Disorders and Diseases
Follicle-stimulating hormone (FSH)	Ovaries and testes **In females:** Stimulates the development of oocytes (egg cells or immature ova) **In males:** Stimulates the production of sperm	**Hyposecretion:** Sterility
Luteinizing hormone (LH)	Ovaries and testes **In females:** Stimulates ovulation, the formation of the corpus luteum, and secretion of estrogens and progesterone **In males:** Stimulates the production of testosterone	**Hyposecretion:** Sterility **Hypersecretion:** Stein-Leventhal syndrome (polycystic ovary syndrome)
Prolactin (PRL), or lactogenic hormone	Mammary glands **In females:** Stimulates the secretion of milk from the breasts **In males:** Action is unknown	**Hypersecretion in females:** Galactorrhea (abnormal lactation), amenorrhea (absence of menstrual cycles)
Adrenocortico-tropic hormone (ACTH), or corticotropin	Adrenal cortex Stimulates and controls the adrenal cortex	**Hyposecretion:** Addison's disease **Hypersecretion:** Cushing's syndrome

Table 7.2: (continued)

Hormone	Target Tissues and Actions	Disorders and Diseases
Melanocyte-stimulating hormone (MSH)	Skin Exact actions are unknown, but can cause darkening of the skin	Well-defined diseases due to hyposecretion or hypersecretion are not yet known.
Posterior pituitary gland The posterior pituitary gland does not synthesize hormones. Instead, it stores and releases hormones synthesized by the hypothalamus		
Oxytocin (OT)	Uterus, mammary glands Stimulates contraction of uterus during labor and stimulates the "milk let-down" reflex during lactation	Well-defined diseases due to hyposecretion or hypersecretion are not yet known.
Antidiuretic hormone (ADH), or vasopressin	Kidneys, sudoriferous glands, blood vessels Antidiuretic effect (i.e., conserves water by decreasing urine volume and perspiration), raises blood pressure	**Hyposecretion:** Diabetes insipidus
Pineal Gland		
Located in the epithalamus (part of the diencephalon) of the brain, where it is attached to the roof of the third ventricle. Made of neuroglia and secretory cells. Produces the hormone melatonin, which is thought to be involved in the sleep/wake cycle (circadian rhythm) as well as the onset of puberty. Melatonin's release is stimulated by darkness and inhibited by light		
Melatonin	Body's biological clock Causes sleepiness	**Hyposecretion:** Insomnia **Hypersecretion:** Seasonal affective disorder (SAD)

Table 7.2: (continued)

Hormone	Target Tissues and Actions	Disorders and Diseases
Thyroid Gland		
Butterfly-shaped gland found wrapped around the trachea just below the larynx. Secretes two hormones that play a vital role in the body's metabolism. These are thyroxine (T_4) and tri-iodothyronine (T_3). T_4 circulates in the body and is converted into T_3, the more biologically active hormone, in the tissues. Together they are often referred to as thyroid hormone. The thyroid also secretes calcitonin		
Thyroid hormone (tri-iodothyronine and thyroxine)	Cells and tissues throughout the body Controls oxygen use and the basal metabolic rate (the minimum amount of energy used by the body to maintain vital processes), cellular metabolism, and growth and development	**Hyposecretion:** Cretinism, myxedema **Hypersecretion:** Graves' disease, thyrotoxicosis **Thyroid enlargement:** Goiter
Calcitonin (CT), or thyrocalcitonin	Bone Lowers blood calcium levels	Well-defined diseases due to hyposecretion or hypersecretion are not yet known.
Parathyroid Gland		
Small, round masses of tissue found on the posterior surfaces of the thyroid gland. They release only one hormone, parathormone. Works together with calcitonin from the thyroid gland and calcitriol from the kidneys to control blood calcium levels		
Parathormone (PTH), or parathyroid hormone	Bone Increases blood calcium and magnesium levels; decreases blood phosphate levels; promotes formation of calcitriol by the kidneys	**Hyposecretion:** Tetany, hypocalcemia **Hypersecretion:** Demineralization of bones

Table 7.2: (continued)

Hormone	Target Tissues and Actions	Disorders and Diseases
Thymus Gland		
Located in the thorax, behind the sternum and between the lungs. Large in infants and reaches its maximum size around puberty, when its size begins to decrease with age. Plays an important role in the immune system and secretes a number of thymic hormones involved in immunity, including thymosin, thymic humoral factor (THF), thymic factor (TF), and thymopoietin		
Thymic hormones	T cells (found throughout the body) Promote growth of T cells, which are a type of white blood cell	Decreased immunity
Pancreatic Islets		
The pancreas is a long organ, approximately 5–6 in. (12.5–15 cm) in length. Found behind and slightly below the stomach and is both an endocrine and an exocrine gland as it also functions in digestion (its digestive functions are discussed in chapter 11). Scattered in the pancreas are small patches of endocrine tissue called pancreatic islets, or islets of Langerhans. Pancreatic islets are composed of four types of hormone-secreting cells: alpha cells, beta cells, delta cells, and PP cells (F cells). These cells secrete four different hormones that generally help regulate blood glucose levels		
Glucagon (from alpha cells)	Liver Accelerates the breakdown of glycogen into glucose; stimulates the release of glucose into the blood—therefore, raises blood glucose levels	**Hypersecretion:** Hyperglycemia
Insulin (from beta cells)	All body cells Accelerates the transport of glucose into cells; converts glucose into glycogen—therefore, lowers blood glucose levels	**Hyposecretion:** Diabetes mellitus **Hypersecretion:** Hyperinsulinism

179

Table 7.2: (continued)

Hormone	Target Tissues and Actions	Disorders and Diseases
Somatostatin (from delta cells)—this is identical to growth-hormone-inhibiting hormone secreted by the hypothalamus	Pancreas Inhibits insulin and glucagon release; slows absorption of nutrients from the gastrointestinal tract	**Hypersecretion:** Diabetes mellitus
Pancreatic polypeptide (from PP cells (F cells))	Pancreas and gall bladder Inhibits secretion of pancreatic fluid, bicarbonate, and enzymes. Inhibits contraction of gall bladder	Well-defined diseases due to hyposecretion or hypersecretion are not yet known.

Adrenal Glands (Suprarenal Glands)

Found above the kidneys. Although an adrenal gland looks like a single organ, it contains two regions that are structurally and functionally different. The outer adrenal cortex surrounds the inner adrenal medulla

Adrenal cortex
Produces steroid hormones that are essential to life, and loss of them can lead to potentially fatal dehydration or electrolyte imbalances. These hormones are grouped into mineralocorticoids, glucocorticoids, and sex hormones

Mineralocorti-coids (mainly **aldosterone**)	Kidneys Regulate mineral content of the blood by increasing blood levels of sodium and water, and decreasing blood levels of potassium	**Hyposecretion of gluco-corticoids and aldoste-rone:** Addison's disease **Hypersecretion of aldosterone:** Aldosteronism
Glucocorticoids (mainly **cortisol**)	All body cells Regulate metabolism; help body resist long-term stressors; control effects of inflammation; depress immune responses	**Hyposecretion of gluco-corticoids and aldoste-rone:** Addison's disease **Hypersecretion:** Cushing's syndrome

Table 7.2: (continued)

Hormone	Target Tissues and Actions	Disorders and Diseases
Sex hormones (**androgens** and **estrogens**)	Very small contribution to sex drive and libido	**Presence of feminizing hormones in males:** Gynecomastia (enlargement of breasts) **Hypersecretion:** Hirsutism

Adrenal medulla
Innervated by neurons of the sympathetic division of the autonomic nervous system and can very quickly release hormones that are collectively referred to as catecholamines. These hormones are, to a large extent, responsible for the fight-or-flight response of the body, and they help the body cope with stress

Adrenaline (epinephrine) and **noradrenaline (norepinephrine)**	All body cells Fight-or-flight response: • Increase blood pressure • Dilate airways to the lungs • Decrease rate of digestion • Increase blood glucose level • Stimulate cellular metabolism	**Hypersecretion:** Prolonged fight-or-flight response

Ovaries and Testes

The female sex glands, the ovaries, and the male sex glands, the testes, are called the gonads. The ovaries are almond-sized organs found in the pelvic cavity. The testes are suspended in a sac, the scrotum, outside the pelvic cavity. The ovaries and testes are discussed in more detail in chapter 13

Ovaries
The ovaries only begin to function properly at puberty when stimulated by FSH and LH from the anterior pituitary gland. The ovaries produce estrogens and progesterone

Table 7.2: (continued)

Hormone	Target Tissues and Actions	Disorders and Diseases
Estrogens (includes estriol, estrone, and estradiol)	Reproductive system Stimulate the development of feminine secondary sex characteristics; together with progesterone, they regulate the female reproductive cycle	Imbalances in secretions can lead to a range of reproductive disorders
Progesterone	Reproductive cycle Together with estrogens, regulates the female reproductive cycle; helps maintain pregnancy	Imbalances in secretions can lead to a range of reproductive disorders

Testes
The testes produce both sperm and male sex hormones, called androgens. The most important androgen is testosterone

Hormone	Target Tissues and Actions	Disorders and Diseases
Testosterone	Reproductive system Stimulates development of masculine secondary sex characteristics promotes growth and maturation of male reproductive system and sperm production; stimulates sex drive	Sterility

Hormones controlled by the pituitary gland

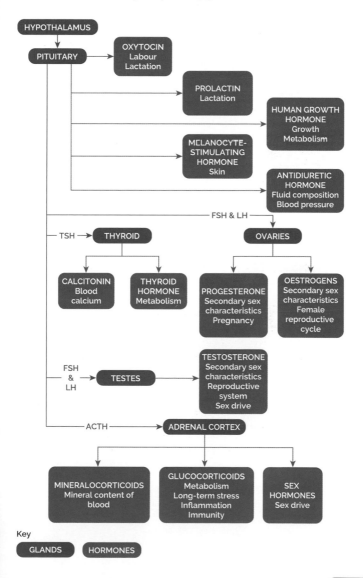

Key

GLANDS   HORMONES

Hormones not controlled by the pituitary gland

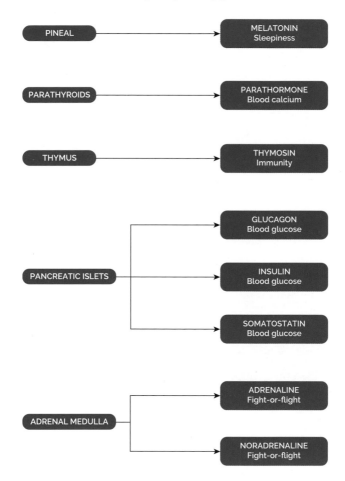

PINEAL → MELATONIN
Sleepiness

PARATHYROIDS → PARATHORMONE
Blood calcium

THYMUS → THYMOSIN
Immunity

PANCREATIC ISLETS →
GLUCAGON
Blood glucose

INSULIN
Blood glucose

SOMATOSTATIN
Blood glucose

ADRENAL MEDULLA →
ADRENALINE
Fight-or-flight

NORADRENALINE
Fight-or-flight

The Stress Response

The stress response involves a series of neuroendocrine reactions that help us survive. These physiological reactions should be only quick and short-lived, but unfortunately many people experience stress on an on-going basis.

The fight-or-flight response, sometimes called the fight-flight-freeze response, is our body's immediate reaction to stressors and it works via the **hypothalamic-pituitary-adrenal axis (HPA axis)**. Nerve impulses from the hypothalamus stimulate the sympathetic division of the autonomic nervous system, which stimulates visceral effectors and the adrenal medulla. Visceral effectors act on cardiac and smooth muscle, while the adrenal medulla releases adrenaline and noradrenaline. Together, these mobilize the fight-or-flight response by inhibiting non-essential body processes, such as digestive, urinary, and reproductive functions, and stimulating the body's resources for fighting or fleeing. Occasionally, a person may freeze in response to a situation.

The resistance reaction is a longer-lasting response to stress and it involves a cascade of hormones whose release is initiated by the hypothalamus. Constantly high levels of stress hormones, especially cortisol, can lead to exhaustion, which often includes hypertension, hyperglycemia, wasting of muscle, and suppression of the immune system.

The Fight-or-Flight Response	
Requirement	Physiology
Increased blood supply to heart, lungs, brain, and skeletal muscles	• Heart rate increases • Blood vessels to heart, lungs, brain, and skeletal muscles dilate • Blood vessels to skin and most viscera contract • Spleen contracts to release stored red blood cells into bloodstream
Increased blood pressure	Reduced blood flow to kidneys activates renin–angiotensin–aldosterone pathway causing kidneys to retain sodium; this elevates blood pressure and preserves blood volume in case of bleeding
Increased oxygen supply	Airways dilate
Increased glucose/ energy supply	Liver converts stored glycogen into glucose

The Resistance Reaction

Hormone Released by Hypothalamus and Acting on Anterior Pituitary Gland	Hormone Released by Anterior Pituitary Gland	Target Organ for Hormone from Anterior Pituitary Gland	Hormone Released by Target Organ	Overall Effect
Corticotropin-releasing hormone (CRH)	Adrenocorticotropic hormone (ACTH)	Adrenal cortex	Cortisol	Increased availability of glucose, fatty acids and amino acids for energy production and cellular repair Reduction in inflammation
Growth-hormone-releasing hormone (GHRH)	Human growth hormone (hGH)	Liver	Insulin-like growth factors (IGFs)	Increased availability of glucose and fatty acids for energy
Thyrotropin-releasing hormone (TRH)	Thyroid-stimulating hormone (TSH)	Thyroid gland	Thyroid hormones (T_3 and T_4)	Increased use of glucose forenergy production

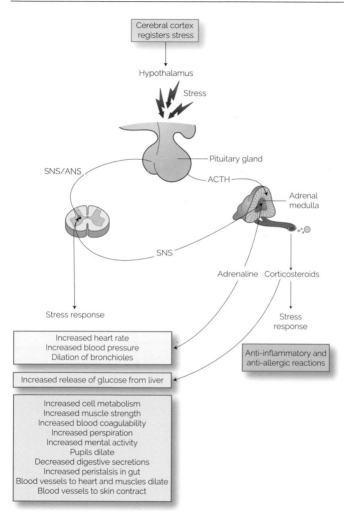

The fight-or-flight response

Figure 7.2: The stress response (ACTH, adrenocorticotropic hormone; ANS, autonomic nervous system; SNS, sympathetic nervous system)

Theory in practice

Surprisingly, our hormones are generally not negatively affected by the process of aging. However, certain changes in hormonal function do mark our journey through life.

Puberty is a significant time in which a child becomes sexually mature. It generally occurs around the age of 12 in girls and 14 in boys and is stimulated by the gonadotropic hormones of the pituitary gland. It is characterized by the appearance of secondary sexual characteristics in both sexes and the start of menstruation in girls.

Pregnancy is a time in which a woman creates a child and becomes a mother. Many hormones are at work in a woman's body during pregnancy, especially estrogen and progesterone. Once properly developed in the uterus, the placenta also secretes hormones to help maintain the pregnancy.

Menopause is a very significant time for women as it marks the end of their ability to bear children. It generally occurs between the ages of 45 and 55 and is characterized by the cessation of ovulation and menstruation.

CHAPTER 8

The Respiratory System

Two systems work closely to ensure there is a continuous supply of oxygen to all the cells of the body and a continuous removal of carbon dioxide. These are the respiratory and cardiovascular systems.

The respiratory system takes in oxygen from the air we breathe and eliminates carbon dioxide, while the cardiovascular system transports these two gases between the respiratory system and the cells of the body.

Functions of the respiratory system include gaseous exchange, olfaction, and sound production.

The lungs have a double blood supply:

- **Pulmonary arteries** bring deoxygenated blood to the lungs. The blood is oxygenated by the lungs and this oxygenated blood is transported from the lungs to the heart by the pulmonary veins. These are the only veins in the body that carry oxygenated blood.
- **Bronchial arteries** bring oxygenated blood to the lung tissue. Most of this blood is returned to the heart by the pulmonary veins. However, some of it drains into the bronchial veins, which transport it to the superior vena cava and then to the heart.

Organization of the Respiratory System

The respiratory system is divided into two zones:

- **The conducting zone:** This is a series of interconnecting passageways that allows air to reach the lungs. No gaseous exchange occurs here. The function of the structures of this zone is to transport air to the alveoli and to filter, moisten, and warm it. Structures of the conducting zone include the nose, pharynx, larynx, trachea, and the bronchi and their smaller branches.
- **The respiratory zone:** This is where the exchange of gases occurs. The alveoli form the respiratory zone.

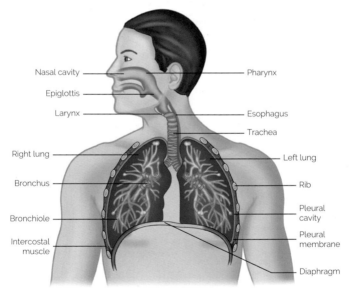

Figure 8.1: Overview of the respiratory system

Table 8.1: Zones and structures of the respiratory system

The Conducting Zone	
Nose	Air enters the system through the nose, where it is filtered, warmed, and moistened. The nose also receives olfactory stimuli and acts as a resonating chamber for sound
Pharynx	Air then goes into the pharynx (throat), which is divided into the nasopharynx, oropharynx, and laryngopharynx. The pharynx is a passageway for air, food, and drink and acts as a resonating chamber for sound
Larynx	From the pharynx air goes into the larynx (voice box), which routes air and food into their correct channels and produces sound. The opening of the larynx is protected by the epiglottis
Trachea	From the larynx, air goes into the trachea (windpipe), which is made of incomplete C-shaped rings of cartilage, and which lies in front of the esophagus. The trachea transports air into the bronchi
Bronchi	The bronchi divide repeatedly to form the bronchial tree
Lungs	The lungs are cone-shaped organs and occupy most of the thoracic cavity. They are separated by the mediastinum and protected by the ribs. The right lung is shorter, thicker, and broader than the left. The lungs are covered and protected by the pleural membrane, which consists of the parietal pleura and visceral pleura
The Respiratory Zone	
Alveoli	Having traveled through the conducting passageways, air arrives at the alveoli (air sacs), where gaseous exchange takes place. The alveoli are surrounded by a dense network of pulmonary capillaries. The respiratory membrane is made up of the thin walls of the alveoli and pulmonary capillaries, and it is the site of gaseous exchange between the lungs and the blood

The nose

The nose is a framework of bone and hyaline cartilage that is covered by skin and lined internally with a mucous membrane. Air enters the nose through the two external nares and flows into the nasal cavity which is divided in two by the septum. At the back of the nasal cavity are two internal nares (choanae), which are openings that connect the nasal cavity to the pharynx.

Air entering the respiratory system through the nose is filtered, warmed, and moistened. The nose also receives olfactory stimuli and acts as a resonating chamber for sound.

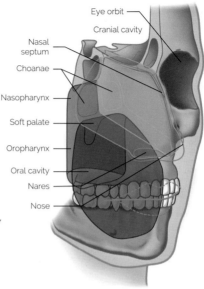

Figure 8.2: Skeleton of the nose

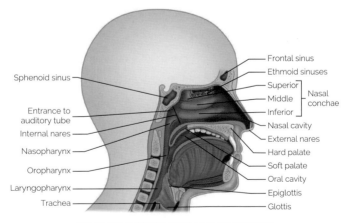

Figure 8.3: Inside the nasal cavity and pharynx

Paranasal sinuses

Paranasal sinuses are air-filled spaces within the cranial and facial bones. They are located near the nasal cavity and serve as resonating chambers for sound when we speak. They are also lined with a mucous membrane and have tiny openings into the nasal cavity called **ostia**.

There are three pairs of paranasal sinuses, named after the bone in which they are located: the frontal, maxillary, and sphenoid sinuses.

There are also the ethmoid sinuses, which consist of many spaces inside the ethmoid bone.

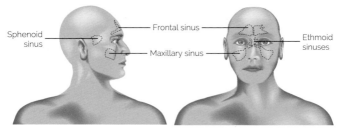

Figure 8.4: Paranasal sinuses

Pharynx (throat)

From the nose, air passes through the internal nares and into the pharynx. This is a funnel-shaped tube whose walls are made up of skeletal muscles lined by mucous membrane and cilia. The mucus traps dust particles, and cilia move the mucus downward. The pharynx is divided into three portions, which are named after the structure to which they are closest: the nasopharynx (nasal cavity), oropharynx (oral cavity), and laryngopharynx or hypopharynx (larynx).

The pharynx is a passageway for air, food, and drink and acts as a resonating chamber for sound.

Larynx (voice box)

The pharynx transports air to the larynx, which is a short passageway between the laryngopharynx and the trachea.

The larynx is made up of eight pieces of rigid hyaline cartilage and a leaf-shaped piece of elastic cartilage called the **epiglottis**, which protects the opening of the larynx. The first piece of cartilage in the larynx is the thyroid cartilage. It gives the larynx its triangular shape. The last piece is the cricoid cartilage, which connects the larynx to the trachea.

The larynx routes air and food into their correct channels and produces sound.

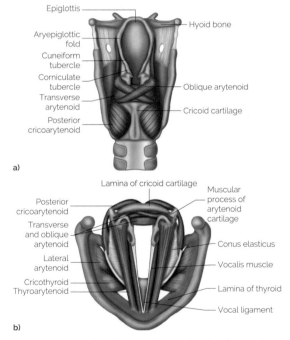

Figure 8.5: Larynx and hyoid bone, (a) posterior view; (b) superior view

Trachea (windpipe)

The trachea is a long, tubular passageway which transports air from the larynx into the bronchi.

The trachea lies in front of the esophagus and is composed of 16–20 incomplete C-shaped rings of hyaline cartilage. The open parts of the C-shape are held together with transverse smooth muscle fibers and elastic connective tissue. This open area lies against the esophagus and allows for expansion of the esophagus during swallowing. The cartilage parts of the C-shape are solid so that they can support the trachea and keep it open despite changes in breathing.

The trachea is lined with mucous membrane and cilia that move any minute dust particles still in the respiratory system upward, away from the lungs to the throat, where they can be swallowed or spat out.

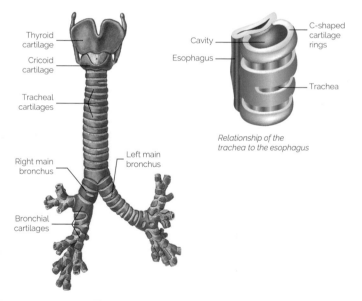

Relationship of the trachea to the esophagus

Figure 8.6: Larynx, trachea, and bronchi

Bronchi

Bronchi are dividing branchlike passageways that carry air from the trachea into the alveoli. They are composed of incomplete rings of cartilage and are lined with mucous membrane.

However—as branching takes place— gradual changes occur to the structure of the branches until the cartilage is finally replaced by spiral bands of smooth muscle and the protective mechanism of the cilia is replaced by the action of macrophages.

Trachea
⇓
Primary bronchi
⇓
Secondary bronchi
⇓
Tertiary bronchi
⇓
Bronchioles
⇓
Terminal bronchioles
⇓
Respiratory bronchioles
⇓
Alveolar ducts

The bronchial tree

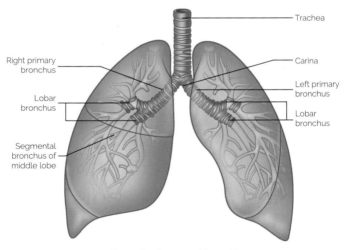

Figure 8.7: Lungs and bronchi

Lungs

The lungs are cone-shaped organs that occupy most of the thoracic cavity. They are protected by the ribs and separated from one another by the heart and the mediastinum, which is a mass of tissue extending from the sternum to the vertebral column. The right lung is shorter, thicker, and broader than the left.

The lungs are covered and protected by the pleural membrane, which consists of the parietal pleura and visceral pleura.

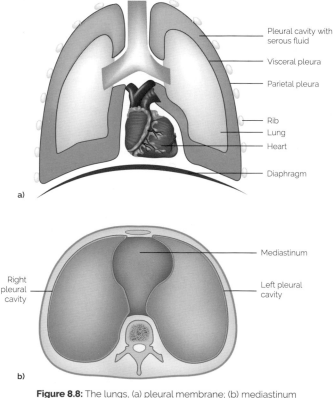

Figure 8.8: The lungs, (a) pleural membrane; (b) mediastinum (transverse section, inferior view)

Alveoli (air sacs)

Alveoli are cup-shaped pouches inside the lungs and are the site of gaseous exchange between the lungs and the blood.

They have extremely thin walls composed mainly of a single layer of squamous epithelial cells and are surrounded by a dense network of pulmonary capillaries. Together the thin walls of the alveoli and pulmonary capillaries—and their basement membranes—form the respiratory membrane where gaseous exchange occurs. It is an extremely thin membrane so that the gases can diffuse rapidly across it.

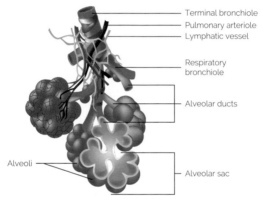

Terminal bronchiole
Pulmonary arteriole
Lymphatic vessel

Respiratory bronchiole

Alveolar ducts

Alveoli

Alveolar sac

Figure 8.9: Portion of a lobule of the lungs

Physiology of Respiration

Respiration is the exchange of gases between the atmosphere, blood, and cells, and it occurs in pulmonary ventilation, external respiration, internal respiration, and cellular respiration. We will now look at pulmonary ventilation and external and internal respiration in more detail.

Pulmonary ventilation

The process of breathing: Air is inspired, or breathed, into the lungs and expired, or breathed, out of the lungs. Pulmonary ventilation is a mechanical process dependent on the existence of a pressure gradient. During inspiration, the diaphragm and external intercostal muscles contract and increase the thoracic cavity. Thus, the volume of the lungs increases, causing a decrease in the pressure in the lungs. When the pressure in the lungs is less than atmospheric pressure, a partial vacuum is created, and air is sucked into the lungs.

Normal expiration is a passive process in which the muscles of the chest and lungs recoil and the volume of the lungs decreases. When the volume of the lungs decreases, the pressure inside the lungs increases and, when it is greater than the atmospheric pressure, air will move out of the lungs to the area of lowest pressure. In active expiration, the abdominal and internal intercostal muscles contract to force the diaphragm upward. This results in a reduction of the size of the thoracic cavity and an increase in the pressure within the lungs. This leads to expiration.

External respiration (pulmonary respiration)

External respiration is the exchange of gases between the alveoli of the lungs and the blood in the pulmonary capillaries. In external respiration, oxygen diffuses from the alveolar air into the blood and carbon dioxide diffuses from the blood into the alveolar air. Oxygen is transported by hemoglobin in the blood. Carbon dioxide is transported as bicarbonate ions in the blood plasma.

Internal respiration (tissue respiration)

Internal respiration is the exchange of gases between the blood and tissue cells. In internal respiration, oxygen diffuses from the blood into the cells and carbon dioxide diffuses from the cells into the blood.

Table 8.2: Composition of inspired and expired air

Gas	Inspired Air (%)	Expired Air (%)
Oxygen	21	16 to 17
Carbon dioxide	0.04	4 to 4.5
Nitrogen	78	78
Inert gases	0.96	0.96
Water vapor	Varies	Varies

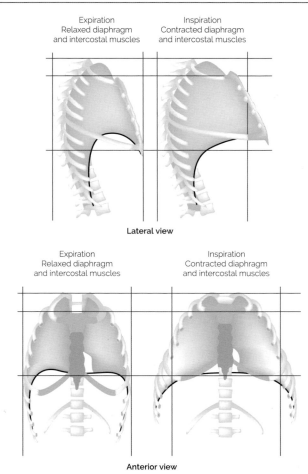

Expiration
Relaxed diaphragm
and intercostal muscles

Inspiration
Contracted diaphragm
and intercostal muscles

Lateral view

Expiration
Relaxed diaphragm
and intercostal muscles

Inspiration
Contracted diaphragm
and intercostal muscles

Anterior view

Figure 8.10: Changes in the thoracic cavity during breathing

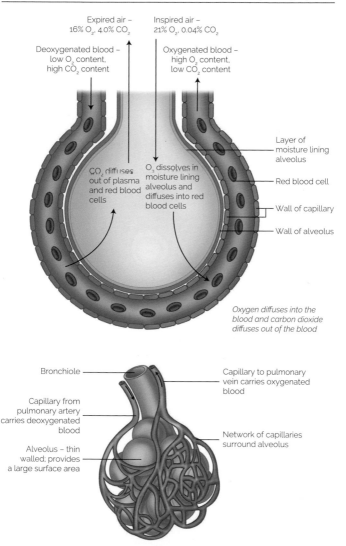

Expired air –
16% O_2, 4.0% CO_2

Inspired air –
21% O_2, 0.04% CO_2

Deoxygenated blood –
low O_2 content,
high CO_2 content

Oxygenated blood –
high O_2 content,
low CO_2 content

CO_2 diffuses
out of plasma
and red blood
cells

O_2 dissolves in
moisture lining
alveolus and
diffuses into red
blood cells

Layer of
moisture lining
alveolus

Red blood cell

Wall of capillary

Wall of alveolus

*Oxygen diffuses into the
blood and carbon dioxide
diffuses out of the blood*

Bronchiole

Capillary to pulmonary
vein carries oxygenated
blood

Capillary from
pulmonary artery
carries deoxygenated
blood

Alveolus – thin
walled; provides
a large surface area

Network of capillaries
surround alveolus

Figure 8.11: Gaseous exchange between the alveoli and capillaries

Theory in practice

The word *lunge* means "lightweight," and although the total surface area of the lungs is approximately 40 times the external surface area of the entire body, they weigh less than a kilogram. These light, "airy" organs work closely with the cardiovascular system to ensure there is a continuous supply of oxygen to all the cells of the body and a continuous removal of carbon dioxide. Every cell in the body needs oxygen to produce energy in the form of adenosine triphosphate (ATP). Without oxygen, cells are unable to produce energy and will therefore die. Oxygen is vital to the survival of cells, yet, ironically, no cells have the ability to store it.

Thankfully, when resting, healthy adults move approximately six liters of air in and out of their lungs every minute and the diaphragm can adapt to our needs: it descends approximately 4/10 in. (1 cm) during normal breathing; however, during strenuous breathing it can descend as much as 4 in. (10 cm).

CHAPTER 9

The Cardiovascular System

Blood

Blood is a liquid connective tissue that is slightly sticky and is heavier, thicker, and more viscous than water. It is composed of blood plasma and formed elements and is a vital substance in the body that functions in transportation, regulation, and protection.

The components of blood are detailed overleaf in table 9.1.

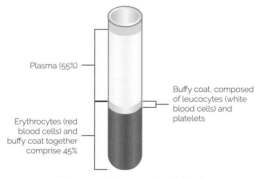

Plasma (55%)

Buffy coat, composed of leucocytes (white blood cells) and platelets

Erythrocytes (red blood cells) and buffy coat together comprise 45%

Figure 9.1: Components of blood

Table 9.1: Components of blood

Components of Blood Plasma		
Component	**Description**	**Function**
Solvent		
Water		• Functions as a **solvent** and suspending medium for carrying other substances • Also absorbs, transports and releases heat
Solutes		
Proteins	Include albumins, globulins, and fibrinogen	• **Albumins** contribute to the osmotic pressure of the blood and transport fatty acids, some lipid-soluble hormones, and certain drugs • **Globulins** transport lipids and function in defense (antibodies are a type of globulin called gamma globulins, or immunoglobulins) • **Fibrinogen** functions in blood clotting
Electrolytes (ions)	Include sodium, potassium, calcium, magnesium, chloride, and bicarbonate	• Function in maintaining osmotic pressure, pH buffering, and regulation of membrane permeability • Also serve as essential minerals
Nutrients	Nutrients from the gastrointestinal tract include amino acids, glucose, fatty acids, and glycerol	Serve as nutrients for the cells
Regulatory substances	Enzymes and hormones	• **Enzymes** catalyze chemical reactions • **Hormones** regulate cellular activity

Table 9.1: (continued)

Components of Blood Plasma		
Component	Description	Function
Solutes (continued)		
Gases	Oxygen, carbon dioxide and nitrogen	• **Oxygen:** Most oxygen is carried in red blood cells, but a minute amount is also transported in the blood plasma • **Carbon dioxide:** Most carbon dioxide is transported in the blood plasma and a small amount is carried by red blood cells • **Nitrogen:** The functions of nitrogen in the blood are not yet known
Wastes	Waste products of metabolism include urea, uric acid, and other substances	Carried to the organs of excretion

Formed Elements		
Component	Description	Function
Erythrocytes (red blood cells) Contain the protein hemoglobin, which transports oxygen in the blood		
Erythrocytes	Biconcave disks with no nucleus and few organelles; full of hemoglobin	Transport oxygen
Leucocytes (white blood cells) Function primarily in protecting the body against foreign microbes and in immune responses. They do not contain hemoglobin and therefore do not have the red color that hemoglobin gives red blood cells. Thus, white blood cells are a pale, "whitish" color		
Most white blood cells live for a few hours to a few days, and they are less numerous than red blood cells (for every white blood cell in the body there are approximately 700 red blood cells). White blood cells can be categorized into granulocytes and agranulocytes		

Formed Elements		
Component	Description	Function

Granulocytes
Granulocytes have multilobed nuclei and contain granules in their cytoplasm. Their names represent the colors of the dyes that they take up when stained in a laboratory

Neutrophils	Cytoplasm has very fine, pale pink granules *(neutro = neutral; takes up both red acid and alkaline methylene blue dyes)*	• Engulf and digest foreign particles and remove waste through the process of **phagocytosis**; thus, they are referred to as **phagocytes** • Phagocytes increase rapidly during infection and are attracted in large numbers to areas of infection
Eosinophils	Cytoplasm has large, red-orange granules *(eosino = red acid dye)*	• Destroy certain parasitic worms, phagocytize antigen-antibody complexes, and combat the effects of some inflammatory chemicals • Eosinophils increase during allergies and infections by parasitic worms
Basophils	Cytoplasm has large, blue-purple granules *(baso = alkaline methylene blue dye)*	• The granules in basophils contain **histamine**, which causes the dilation of blood vessels • Basophils release histamine at sites of inflammation and are closely associated with allergic reactions

Agranulocytes
Agranulocytes have a large nucleus and do not contain cytoplasmic granules

Lymphocytes	Include T cells, B cells, and natural killer cells	Lymphocytes play an important role in the immune response and are present in lymphatic tissue, such as lymph nodes and the spleen (immunity is discussed in more detail in chapter 10)

Formed Elements		
Component	**Description**	**Function**
Monocytes	The largest white blood cells	• Some monocytes circulate in the blood and are phagocytes • Other monocytes migrate into the tissues and become **macrophages**, which are large scavenging cells that clean up areas of infection • Monocytes increase in number during chronic infections
Thrombocytes (platelets)	Granular, disk-shaped cell fragments containing no nucleus	• Function in **hemostasis**, which is the process by which bleeding is stopped • Thrombocytes form a platelet plug and release chemicals that promote blood clotting

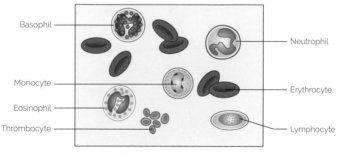

Figure 9.2: Blood cells

Heart

The heart is located in the mediastinum between the lungs in the thoracic cavity and is covered and protected by a triple-layered sac called the pericardium. The pericardium is composed of the outer fibrous pericardium and the inner serous pericardium, which forms a double layer (the outer parietal layer and the inner visceral layer). The walls of the heart are composed of three layers: the outer epicardium, the middle myocardium, and the inner endocardium. The blood supply to the heart tissue is called the coronary circulation (cardiac circulation).

The heart is a hollow, muscular organ divided into left and right sides. Each side is composed of two chambers: an **atrium** for receiving blood into the heart and a **ventricle** for pumping blood out of the heart. Blood is prevented from flowing backward by atrioventricular valves (AV), which separate the atria and ventricles, and semilunar valves, which separate the ventricles and arteries.

The heart is a double pump, which pumps blood into two different circulations:

- Pulmonary circulation: The right side of the heart receives deoxygenated blood from the body and pumps it to the lungs where it is oxygenated.
- Systemic circulation: The left side of the heart receives oxygenated blood from the lungs and pumps it to the rest of the body.

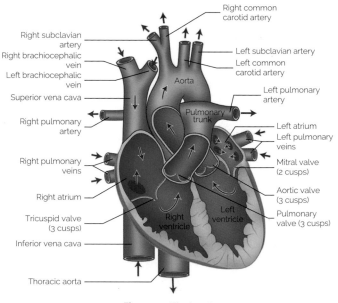

Figure 9.3: The heart

Walls, chambers, and valves of the heart

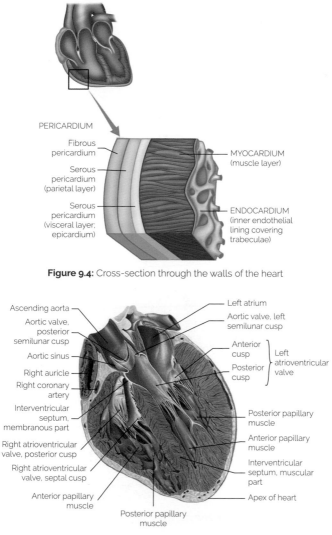

PERICARDIUM

Fibrous pericardium

Serous pericardium (parietal layer)

Serous pericardium (visceral layer; epicardium)

MYOCARDIUM (muscle layer)

ENDOCARDIUM (inner endothelial lining covering trabeculae)

Figure 9.4: Cross-section through the walls of the heart

Ascending aorta

Aortic valve, posterior semilunar cusp

Aortic sinus

Right auricle

Right coronary artery

Interventricular septum, membranous part

Right atrioventricular valve, posterior cusp

Right atrioventricular valve, septal cusp

Anterior papillary muscle

Posterior papillary muscle

Left atrium

Aortic valve, left semilunar cusp

Anterior cusp

Posterior cusp

Left atrioventricular valve

Posterior papillary muscle

Anterior papillary muscle

Interventricular septum, muscular part

Apex of heart

Figure 9.5: Chambers of the heart

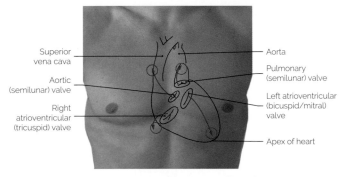

Figure 9.6: Contour of the heart and its valves

Coronary circulation: The blood supply to the heart

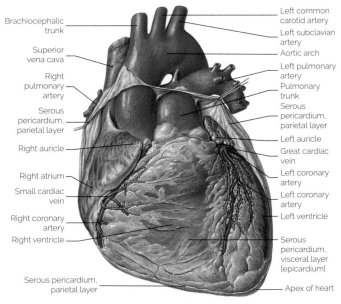

Figure 9.7: Coronary blood vessels

Blood flow to the heart tissue

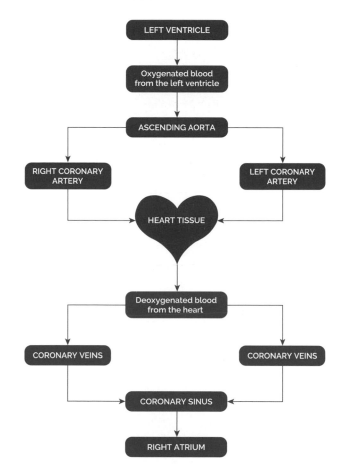

A small amount of deoxygenated blood from the heart tissue filters directly into the heart chambers via venous channels.

Key

DEOXYGENATED BLOOD   OXYGENATED BLOOD

Pulmonary and systemic circulation: The flow of blood through the heart

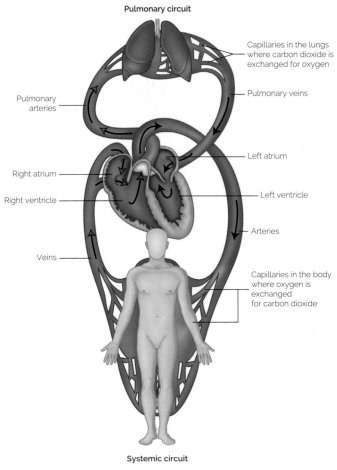

Pulmonary circuit

Capillaries in the lungs where carbon dioxide is exchanged for oxygen

Pulmonary veins

Pulmonary arteries

Left atrium

Right atrium

Left ventricle

Right ventricle

Arteries

Veins

Capillaries in the body where oxygen is exchanged for carbon dioxide

Systemic circuit

Figure 9.8: Pulmonary and systemic circulation

Blood flow through the heart

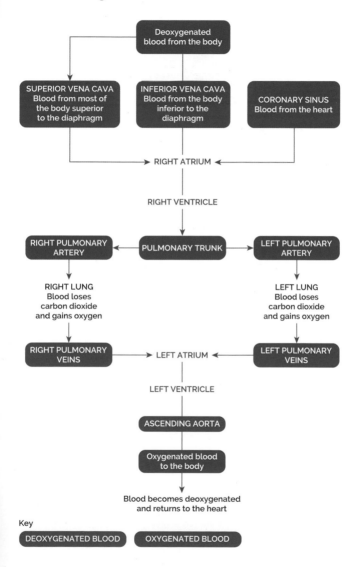

Key

DEOXYGENATED BLOOD   OXYGENATED BLOOD

Physiology of the heart

The cardiac cycle is all the events associated with a heartbeat. Starting from the end of the previous heartbeat, a cardiac cycle involves cardiac diastole (relaxation), in which the chambers begin to fill with blood; atrial systole (contraction), in which the atria contract and empty their contents into the ventricles; ventricular systole, in which the ventricles contract and pump blood into the arteries; and then the cycle returns to cardiac diastole.

The heart rate is the number of times the heart beats in one minute, and it can be modified by the autonomic nervous system: sympathetic stimulation increases the heart rate and parasympathetic stimulation decreases the heart rate.

Resting heart rate is normally 60–100 bpm. Cardiac muscle cells have a myogenic rhythm and contract independently of nervous stimulation, and the rhythm of the heart is mainly controlled by the intrinsic conduction system (nodal system). The heart rate can also be modified by chemicals such as hormones, ions, certain drugs, and gases.

Pacemaker

SINOATRIAL NODE
(SA node)
Function: Initiates impulses and conducts them throughout both atria.
Location: right atrial wall.

ATRIOVENTRICULAR NODE
(AV node)
Function: Receive impulses from SA node and passes them to AV bundle of His. Impulses are slightly delayed here to give the atria time to finish contracting.
Location: atrial septum.

ATRIOVENTRICULAR BUNDLE OF HIS
(AV bundle of His)
Function: Is a bundle of fibres that acts as the electrical connection between the atria and ventricles. Receives impulses from AV node and passes them to right and left bundle branches.
Location: interventricular septum.

Blood is ejected upwards

PURKINJE FIBRES
(Conduction myofibres)
Function: Receive impulses from the bundle branches and conducts them to the ventricular myocardium. Conduct impulses to the apex of the heart first and then upwards to the rest of the heart so that the blood is ejected upwards into the arteries.
Location: ventricular myocardium

RIGHT AND LEFT BUNDLE BRANCHES
Function: Conduct impulses towards the apex of the heart. Receive impulses from AV bundle of His and conduct them to Purkinje fibres.
Location: interventricular septum towards the apex of the heart.

Apex of heart

The intrinsic conduction system

Blood Vessels

Table 9.2: Blood vessels

Arteries and arterioles	Carry blood away from the heart. They are composed of a lumen surrounded by the tunica intima (interna), then the tunica media, and finally the outer tunica adventitia (tunica externa). Elastic (conducting) arteries are large arteries that conduct blood from the heart to the medium-sized arteries. Muscular (distributing) arteries are medium-sized arteries that distribute blood to various parts of the body. Arterioles are tiny arteries that deliver blood to capillaries
Capillaries	Branch throughout tissues and are the site for the exchange of nutrients, oxygen, and waste between the blood and interstitial fluid. The walls of capillaries consist of a single layer of endothelium and a basement membrane. Capillaries are extremely thin to allow for the rapid exchange of substances
Veins and venules	Carry blood away from tissues toward the heart. The structure of veins is similar to that of arteries, except that their walls are thinner and their lumen is larger. Some veins have valves to prevent backflow of blood. Blood is moved through veins by the milking action of skeletal muscles and the movement of the diaphragm

Table 9.3: Differences between arteries and veins

Features	Arteries	Veins
Structure	• Thick walls • No valves • Smaller lumen	• Thin walls • Valves • Larger lumen
Function	Carry blood away from the heart	Carry blood toward the heart
Blood pressure	Higher	Lower

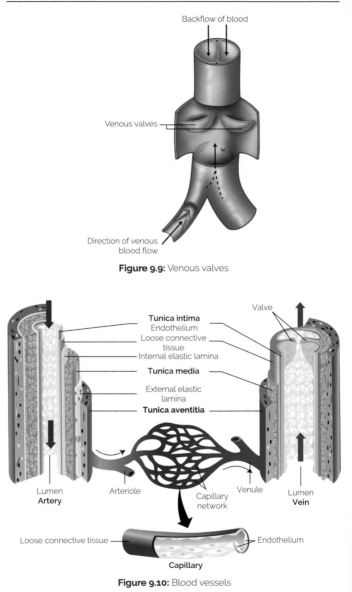

Figure 9.9: Venous valves

Figure 9.10: Blood vessels

Table 9.4: Principal blood vessels of the heart

Vessel	Blood	From	To
Arteries Arteries carry blood **away** from the heart and generally carry **oxygenated** blood			
Aorta	Oxygenated	Heart	Most of the body
Coronary artery	Oxygenated	Heart	Heart tissue
Pulmonary artery	Deoxygenated	Heart	Lungs
Note: The pulmonary artery is the only artery in the body that transports deoxygenated blood			
Veins Veins carry blood **toward** the heart and generally carry **deoxygenated** blood			
Superior vena cava	Deoxygenated	Most of the body superior to the diaphragm, except the alveoli and heart	Heart
Inferior vena cava	Deoxygenated	Body inferior to the diaphragm	Heart
Coronary sinus	Deoxygenated	Heart tissue	Heart
Pulmonary vein	Oxygenated	Lungs	Heart
Note: The pulmonary vein is the only vein in the body that transports oxygenated blood			

Blood Pressure

Blood pressure is the force exerted by blood on the walls of a vessel. It is highest in the aorta and lowest in the inferior vena cava, and in a healthy adult it is approximately 120/80 mm Hg. The first figure is the systolic pressure and the second is the diastolic pressure.

Blood pressure is affected by cardiac output and resistance. Cardiac output is affected by the heart rate, and resistance is affected by changes in the tunica media of arterioles, blood viscosity, and blood vessel length and radius.

Primary Blood Vessels of Systemic Circulation

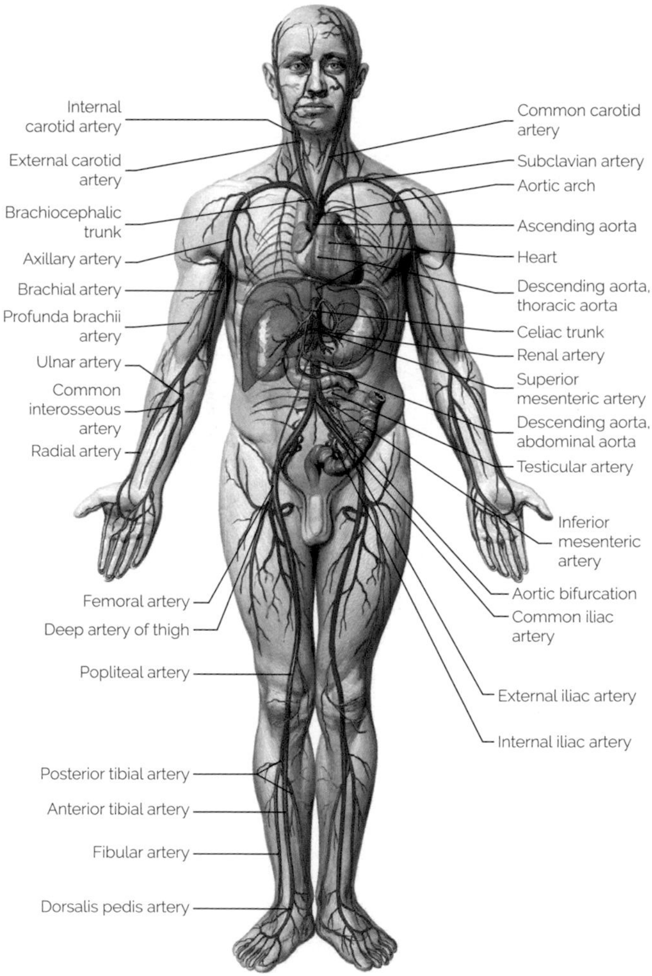

Internal carotid artery
External carotid artery
Brachiocephalic trunk
Axillary artery
Brachial artery
Profunda brachii artery
Ulnar artery
Common interosseous artery
Radial artery

Common carotid artery
Subclavian artery
Aortic arch
Ascending aorta
Heart
Descending aorta, thoracic aorta
Celiac trunk
Renal artery
Superior mesenteric artery
Descending aorta, abdominal aorta
Testicular artery
Inferior mesenteric artery
Aortic bifurcation
Common iliac artery
External iliac artery
Internal iliac artery

Femoral artery
Deep artery of thigh
Popliteal artery
Posterior tibial artery
Anterior tibial artery
Fibular artery
Dorsalis pedis artery

Figure 9.11: Arteries of the systemic circulation

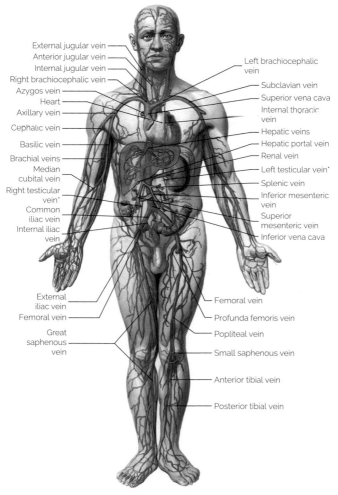

Figure 9.12: Veins of the systemic circulation

Primary arteries of the head, face, and neck

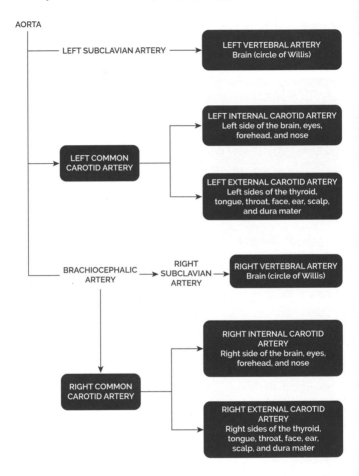

Primary veins of the head, face, and neck

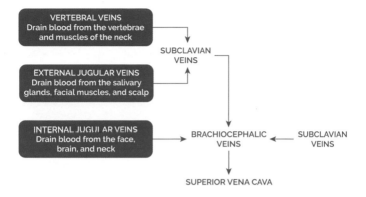

Guide to flowcharts of the blood vessels of the body

Some of the primary blood vessels of the body have been described in flow charts on the following pages. To help you understand these charts please note the following features of the charts:

Colour: The colour of the boxes and arrows represents the type of blood transported. Arteries carrying **oxygenated** blood are coloured red. Veins carrying **deoxygenated** blood are coloured blue.

Coloured boxes: The primary vessels associated with an area are in coloured boxes. These are the vessels you will need to learn for that particular area. The other vessels are simply to help you picture the flow of the blood.

Direction of arrows: The direction of the arrows represents the direction of blood flow. In general, blood flowing in arteries flows downward from the heart to the rest of the body and blood flowing in veins moves upward from the rest of the body to the heart.

Right/left side representation: Most blood vessels of the body are the same on both sides of the body. Thus, for ease of learning, the left and right sides are not noted. However, if there are differences between the left and right sides of the body, then both sides are represented.

Primary arteries of the upper limbs

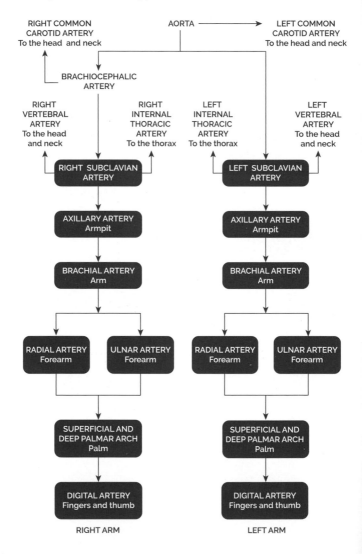

Primary veins of the upper limbs

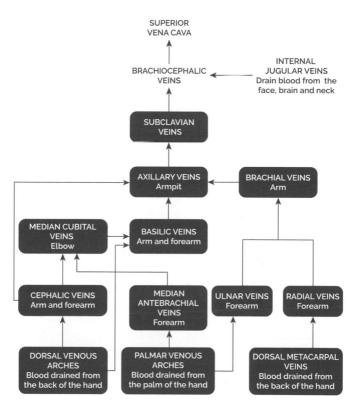

Primary arteries of the thorax

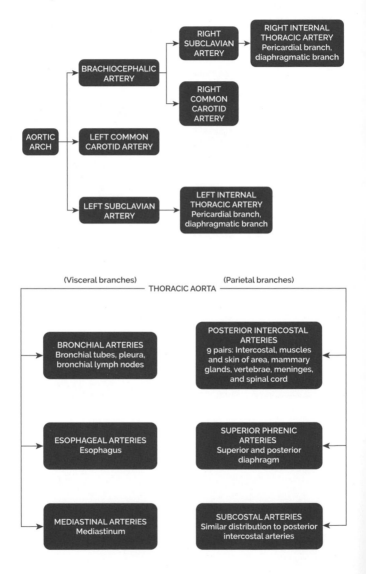

Primary veins of the thorax

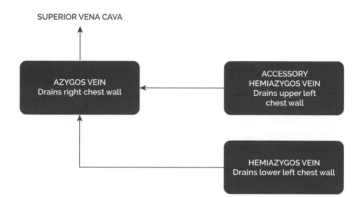

Primary arteries of the abdomen

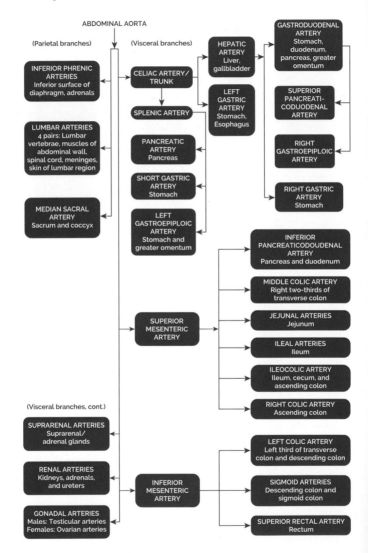

ABDOMINAL AORTA

(Parietal branches)

(Visceral branches)

INFERIOR PHRENIC ARTERIES
Inferior surface of diaphragm, adrenals

CELIAC ARTERY/ TRUNK

HEPATIC ARTERY
Liver, gallbladder

GASTRODUODENAL ARTERY
Stomach, duodenum, pancreas, greater omentum

SPLENIC ARTERY

LEFT GASTRIC ARTERY
Stomach, Esophagus

SUPERIOR PANCREATI-CODUODENAL ARTERY

LUMBAR ARTERIES
4 pairs: Lumbar vertebrae, muscles of abdominal wall, spinal cord, meninges, skin of lumbar region

PANCREATIC ARTERY
Pancreas

RIGHT GASTROEPIPLOIC ARTERY

SHORT GASTRIC ARTERY
Stomach

MEDIAN SACRAL ARTERY
Sacrum and coccyx

LEFT GASTROEPIPLOIC ARTERY
Stomach and greater omentum

RIGHT GASTRIC ARTERY
Stomach

SUPERIOR MESENTERIC ARTERY

INFERIOR PANCREATICODOUDENAL ARTERY
Pancreas and duodenum

MIDDLE COLIC ARTERY
Right two-thirds of transverse colon

JEJUNAL ARTERIES
Jejunum

ILEAL ARTERIES
Ileum

ILEOCOLIC ARTERY
Ileum, cecum, and ascending colon

RIGHT COLIC ARTERY
Ascending colon

(Visceral branches, cont.)

SUPRARENAL ARTERIES
Suprarenal/ adrenal glands

LEFT COLIC ARTERY
Left third of transverse colon and descending colon

RENAL ARTERIES
Kidneys, adrenals, and ureters

INFERIOR MESENTERIC ARTERY

SIGMOID ARTERIES
Descending colon and sigmoid colon

GONADAL ARTERIES
Males: Testicular arteries
Females: Ovarian arteries

SUPERIOR RECTAL ARTERY
Rectum

Primary veins of the abdomen

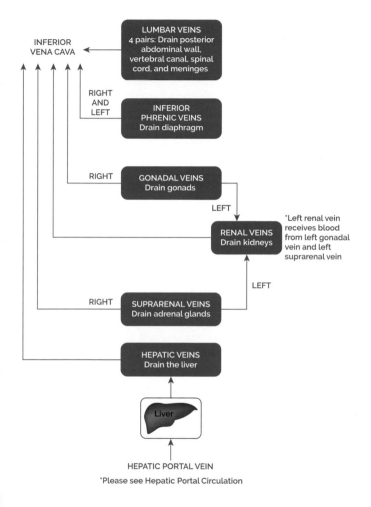

INFERIOR
VENA CAVA

LUMBAR VEINS
4 pairs: Drain posterior
abdominal wall,
vertebral canal, spinal
cord, and meninges

RIGHT
AND
LEFT

**INFERIOR
PHRENIC VEINS**
Drain diaphragm

RIGHT

GONADAL VEINS
Drain gonads

LEFT

RENAL VEINS
Drain kidneys

*Left renal vein
receives blood
from left gonadal
vein and left
suprarenal vein

LEFT

RIGHT

SUPRARENAL VEINS
Drain adrenal glands

HEPATIC VEINS
Drain the liver

Liver

HEPATIC PORTAL VEIN

*Please see Hepatic Portal Circulation

Hepatic portal circulation

Blood from tissues usually flows through one capillary bed before it is returned to the heart. However, blood from the digestive organs passes through a second capillary bed at the liver before it is returned to the heart. This is necessary because blood from the digestive organs is rich in absorbed nutrients but may also contain harmful substances.

The liver stores or modifies the nutrients in order to maintain correct nutrient concentrations in the blood, and it also detoxifies the blood to ensure any harmful substances are not transported to the rest of the body.

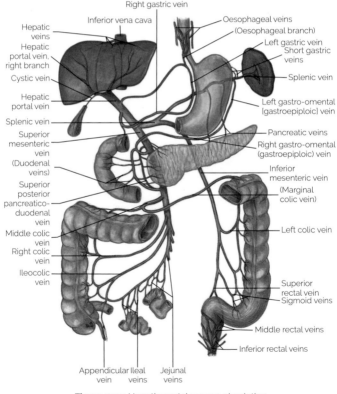

Figure 9.13: Hepatic portal venous circulation

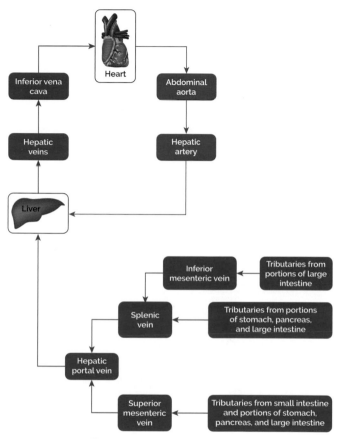

Figure 9.14: Hepatic portal circulation

Primary arteries of the pelvis and lower limbs

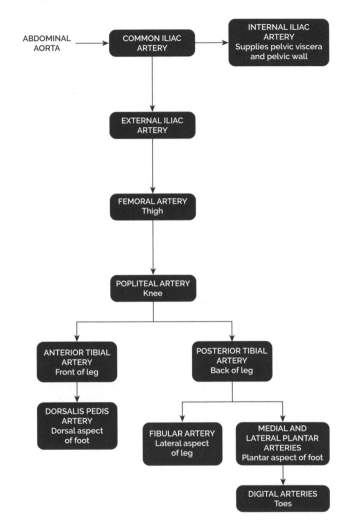

ABDOMINAL AORTA → COMMON ILIAC ARTERY

INTERNAL ILIAC ARTERY
Supplies pelvic viscera and pelvic wall

EXTERNAL ILIAC ARTERY

FEMORAL ARTERY
Thigh

POPLITEAL ARTERY
Knee

ANTERIOR TIBIAL ARTERY
Front of leg

POSTERIOR TIBIAL ARTERY
Back of leg

DORSALIS PEDIS ARTERY
Dorsal aspect of foot

FIBULAR ARTERY
Lateral aspect of leg

MEDIAL AND LATERAL PLANTAR ARTERIES
Plantar aspect of foot

DIGITAL ARTERIES
Toes

Primary veins of the pelvis and lower limbs

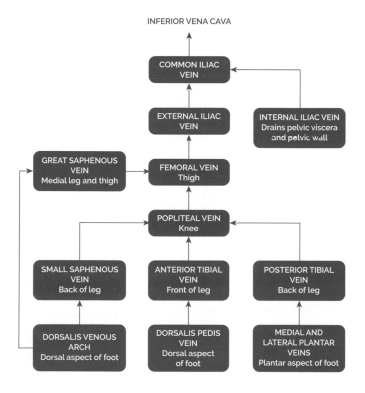

INFERIOR VENA CAVA

COMMON ILIAC VEIN

EXTERNAL ILIAC VEIN

INTERNAL ILIAC VEIN
Drains pelvic viscera and pelvic wall

GREAT SAPHENOUS VEIN
Medial leg and thigh

FEMORAL VEIN
Thigh

POPLITEAL VEIN
Knee

SMALL SAPHENOUS VEIN
Back of leg

ANTERIOR TIBIAL VEIN
Front of leg

POSTERIOR TIBIAL VEIN
Back of leg

DORSALIS VENOUS ARCH
Dorsal aspect of foot

DORSALIS PEDIS VEIN
Dorsal aspect of foot

MEDIAL AND LATERAL PLANTAR VEINS
Plantar aspect of foot

Theory in practice

We have approximately 100,000 km (60,000 miles) of vessels transporting blood to and from over 60 billion cells. When a blood vessel is injured there needs to be a quick, localized response to stop the bleeding before there is excessive blood loss. This process is called hemostasis, and it occurs in three phases:

- **Vascular spasm:** If damaged, the smooth muscles in the wall of a blood vessel immediately contract to narrow the vessel and thus reduce blood flow through it. This contraction is thought to be caused by reflexes initiated by pain receptors in the vessels as well as by serotonin, which is released by platelets once they adhere to the damaged site.
- **Platelet plug formation:** When a blood vessel is damaged, collagen fibers that usually lie under the endothelial cells are exposed and platelets are able to adhere to them. The platelets clump together and release chemicals that attract more platelets to the site. Soon, a mass of platelets called a platelet plug, or white thrombus, forms and seals the injury.
- **Coagulation (blood clotting):** In addition, blood begins to clot or thicken and form a gel at the site of injury. Blood clotting is a complex process, which is promoted by the release of chemicals from platelets. This process results in a mesh of fibrin protein fibers in which blood cells are trapped.

CHAPTER 10

The Lymphatic and Immune System

The lymphatic system drains interstitial fluid to prevent tissues from becoming waterlogged, transports dietary lipids, and protects the body against invasion. Lymphatic vessels carry lymph from capillaries through lymph nodes, into lymphatic trunks, and then into lymphatic ducts. These ducts empty their contents into the left and right subclavian veins. The lymphatic system works closely with the cardiovascular system to maintain health.

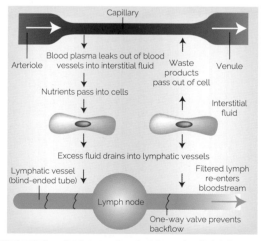

The lymphatic system is a secondary circulatory system that drains excess fluids from tissues and returns the fluid to the cardiovascular system.

Figure 10.1: Connection between blood and lymph

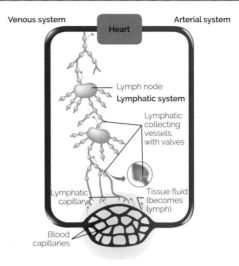

Figure 10.2: Relationship between the lymphatic and cardiovascular systems

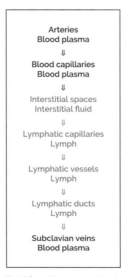

Fluid flow through the body

Organization of the Lymphatic and Immune System

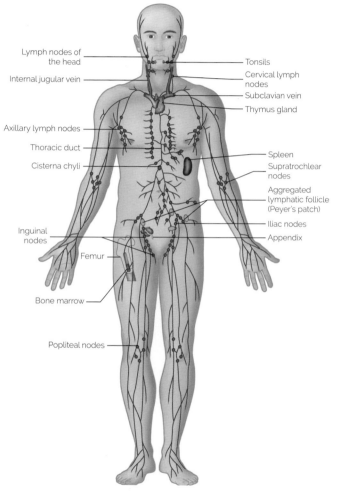

Figure 10.3: Overview of the lymphatic and immune system

Table 10.1: Components of the lymphatic system

Lymph	A clear, straw-colored fluid derived from interstitial fluid and found only in lymphatic vessels. It contains protein molecules, lipid molecules, foreign particles, cell debris, and lymphocytes
Lymphatic capillaries	Closed-ended vessels that permit fluid to flow into them but not out of them. They absorb lymph and transport it to lymphatic vessels
Lymphatic vessels	Carry lymph from capillaries, through nodes, and into lymphatic trunks. Lymph is moved through the vessels by smooth muscle, skeletal muscle, and breathing movements and is prevented from flowing backward by one-way valves
Lymphatic trunks	Trunks are named after the areas they serve and are the: • Lumbar trunk • Intestinal trunk • Right and left bronchomediastinal trunks • Right and left subclavian trunks • Right and left jugular trunks They empty lymph into the thoracic duct and the right lymphatic duct
Lymphatic ducts	There are two lymphatic ducts: • The thoracic duct drains lymph from the left side of the head, neck, and chest, as well as the left arm and the entire body below the ribs. It originates near the second lumbar vertebra at a dilation called the cisterna chyli and empties its contents into the left subclavian vein • The right lymphatic duct drains lymph from the upper right side of the body only and empties into the right subclavian vein
Lymph nodes	Filter the lymph and remove or destroy any potentially harmful substances before the lymph is returned to the blood. They also produce lymphocytes, which function in the immune response

Table 10.1: (continued)

Lymphatic organs	Lymphatic organs include the thymus gland and the spleen:
	
	• The thymus gland is located in the mediastinum, behind the sternum and between the lungs. Not all the functions of the thymus gland are yet known; however, it is known that the gland produces hormones such as thymosin that help the development and maturation of T cells
	• The spleen is the largest single mass of lymphatic tissue in the body. It is located in the abdomen, behind and to the left of the stomach. The spleen filters blood, destroys worn-out red blood cells, stores platelets and blood, and produces lymphocytes. The functional part of the spleen consists of two different tissues:
	  1. white pulp, which functions in immunity and is the site of antibody-producing plasma cells
	  2. red pulp, which functions in the phagocytosis of bacteria, red blood cells, and platelets

Lymphatic nodules, or mucosa-associated lymphoid tissue (MALT)	These are concentrations of lymphatic tissue that are strategically positioned to help protect the body from pathogens that have been inhaled or digested or have entered the body via external openings. They are scattered throughout the mucous membranes that line systems exposed to the external environment, and include the tonsils, Peyer's patches, and the appendix

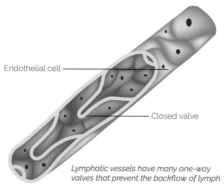

Endothelial cell

Closed valve

Lymphatic vessels have many one-way valves that prevent the backflow of lymph.

Figure 10.4: A lymphatic vessel

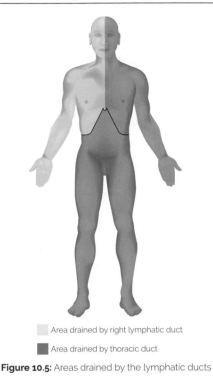

Area drained by right lymphatic duct

Area drained by thoracic duct

Figure 10.5: Areas drained by the lymphatic ducts

Lymph nodes

Lymph nodes are specially structured to allow lymph to flow slowly through them and be filtered of potentially harmful particles and substances. They are scattered along the length of the lymphatic vessels, with higher concentrations of them being strategically placed in sites where there is a greater risk of infection.

Lymph flows very slowly through the nodes and as it flows any foreign particles or substances in it are trapped by the reticular fibers of the nodes. Macrophages, antibodies, and lymphocytes then all work to protect the body against these foreign substances.

Table 10.2: Primary lymph nodes

Name	Location
Lymph Nodes of the Head and Neck	
Superficial parotid nodes (anterior auricular)	In front of ears
Mastoid nodes (posterior auricular)	Behind ears
Submandibular nodes	Beneath mandible
Submental nodes	Beneath chin
Occipital nodes	Base of skull
Deep cervical nodes	Deep within neck
Superficial cervical nodes (includes medial and lateral nodes)	Side of neck
Lymph Nodes of the Body	
Axillary nodes	Armpit
Supratrochlear nodes	Elbow crease
Ileocolic nodes	Abdomen (near the diaphragm)
Iliac nodes	Abdomen
Inguinal nodes	Groin
Popliteal nodes	Knee

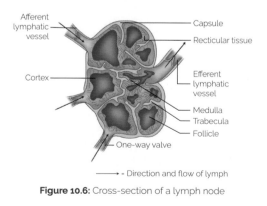

Figure 10.6: Cross-section of a lymph node

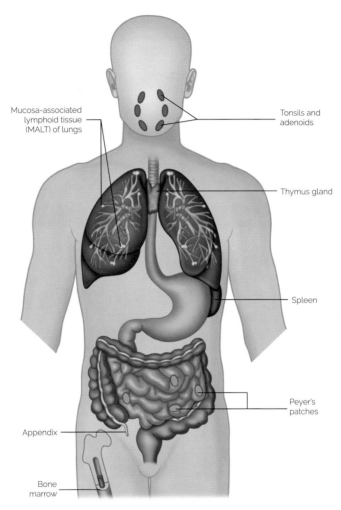

Figure 10.7: Lymphatic organs and nodules

Resistance to Disease and Immunity

The body has a number of different defense mechanisms that are in place to ward off invading microbes and help the body deal with any microbes that actually enter into the system. These mechanisms give general protection against invaders and are classified as non-specific mechanisms. Defense mechanisms in which the body produces specific responses to particular organisms fall under the classification of immunity, and this involves the body's ability to recognize microbes, respond to them appropriately, and remember them so that a secondary response will be quicker and more powerful.

Non-specific resistance to disease

Table 10.3: Non-specific resistance to disease

Mechanical barriers	These are the body's first line of resistance to invaders and include the skin, mucous membranes, tears, saliva, and the processes of urination, defecation, and vomiting
Chemical barriers	These include lysozyme, interferons, the complement system, sebum, perspiration, gastric juice, and vaginal secretions
Natural killer cells	A type of lymphocyte that can kill a variety of microbes as well as some tumor cells. It is not yet known exactly how they recognize or destroy their targets, yet they are found in the spleen, lymph nodes, red bone marrow, and blood
Phagocytes	A type of lymphocyte that ingests microbes and foreign matter. There are two main types of phagocyte: Neutrophils and macrophages
Inflammation	This is the body's response to injury and is a defensive reaction whose purpose is to help prevent the spread of further damage, to prepare the site for repair, and to help clear the area of microbes and toxins. Inflammation has four key signs and symptoms: Swelling, heat, redness, and pain. Occasionally, loss of function of the affected area can be a fifth sign
Fever	This is an abnormally high temperature and is the body's way of dealing with infection. Toxins released by microbes can increase the body's temperature, and this increase in temperature intensifies the effects of the body's own antimicrobial substances, inhibits microbial growth, and increases the speed of tissue repair

Table 10.4: The process of inflammation

Vasodilation and increased permeability of blood vessels	Immediately after tissue damage, blood vessels vasodilate and become more permeable. This is partly caused by the release of histamine. Vasodilation results in more blood flowing to the damaged area and more defensive materials—such as antibodies, phagocytes, and clot-forming chemicals—leaving the blood and entering the injured site. This flow of blood to the area causes swelling, heat, and redness. The pain that usually accompanies inflammation is caused by either injury or irritation to nerve fibers. It can also result from increased pressure caused by the swelling
Phagocyte migration and tissue repair	About an hour after the inflammatory process has begun, large numbers of phagocytes leave the bloodstream and enter the injured site. Neutrophils are the first type of phagocyte to appear. They engulf microbes and foreign material but soon exhaust themselves and rapidly die off. They are then followed by wandering macrophages, which engulf the damaged tissue, invading microbes, and worn-out neutrophils. As the tissue repairs, pus forms. This is a collection of dead cells and fluid. Pus formation continues until the infection has subsided

Immunity (the immune response)

Any substance that the body recognizes as foreign and that provokes a response from the immune system is called an antigen, and the lymphocytes responsible for responding to antigens are referred to as immunocompetent cells. There are two types of immune responses: Cell-mediated immune responses and antibody-mediated immune responses. Pathogens can provoke either one type of response or both types at the same time. The body also has the ability to remember and recognize antigens that have previously caused an immune response. This is known as acquired immunity or immunological memory.

Table 10.5: Immunocompetent cells

B cells	Develop and mature in red bone marrow throughout life. They develop into plasma cells and are able to synthesize and secrete antibodies
T cells	Develop in red bone marrow and then migrate to the thymus gland, where they mature. Two types of T cells exist: CD4+ cells and CD8+ cells. Before T cells are released into the system, they acquire distinctive surface proteins that are capable of recognizing specific antigens and are called antigen receptors

Table 10.6: Types of immune responses

Cell-mediated (cellular) immune responses (CMI responses)	Cells attack antigens directly. CD8+ T cells reproduce into "killer cells," which leave lymphatic tissues to seek out and destroy antigens. This is the most common response to intracellular pathogens such as fungi, parasites, and viruses, as well as some cancer cells and tissue or organ transplants
Antibody-mediated (humoral) immune responses (AMI responses)	Antibodies bind to antigens and inactivate them. B cells develop into plasma cells, which secrete antibodies. Antibodies then leave the lymphatic tissue to circulate in the blood and lymph and bind to the particular antigen for which they were made. Once bound to this antigen they destroy it. This response is more common against antigens that are dissolved in body fluids and pathogens such as bacteria that have multiplied in body fluids
Note: CD4+ T cells become "helper" T cells that aid both CMI and AMI responses	

Table 10.7: Types of immunity

Type of Immunity	How is it Acquired?
Naturally Acquired Immunity	
Naturally acquired active immunity	The body is stimulated to produce its own antibodies through actively having the disease
Naturally acquired passive immunity	This is the transference of antibodies from mother to fetus across the placenta or from mother to baby through breastfeeding
Artificially Acquired Immunity	
Artificially acquired active immunity	Antigens are introduced to the body in the form of vaccinations that induce an active immune response but do not make the recipient ill
Artificially acquired passive immunity	Ready-made antibodies are injected into the recipient

Theory in practice

Sometimes our immune systems can "overreact." For example, an allergy is an overreaction to a substance that is normally harmless to most people. Any substance that invokes an allergic reaction is called an allergen, and foods such as milk, peanuts, shellfish, and eggs are common allergens, as are some antibiotics, vitamins, drugs, venoms, cosmetics, plant chemicals, pollens, dust, and molds. Allergic reactions can only occur if a person has been previously exposed to the allergen and so developed antibodies to it.

Symptoms of allergic reactions can range from a running nose and streaming eyes to anaphylactic shock, in which there is swelling, heart and lung failure, and possibly death. This is because histamine plays a major role in allergies. When histamine is released into the bloodstream it causes the contraction of smooth muscle fibers in the lungs and gastrointestinal tract as well as vasodilation, increased permeability, and increased protein leakage of blood vessels, resulting in low blood pressure, swelling, itching, and redness.

CHAPTER 11

The Digestive System

The **gastrointestinal (GI) tract**, or **alimentary canal**, is a tube that runs from the mouth to the anus. Functions of the digestive system include **ingestion**, **secretion**, **mixing and propulsion**, **digestion**, **absorption**, and **defecation**. Digestion is the process by which large molecules of food are broken down into smaller molecules, either mechanically or chemically, so that they can enter cells. Chemical digestion of food relies on the presence of enzymes, which are catalysts that speed up reactions but do not actually become involved in the reactions themselves.

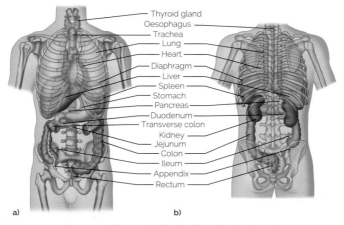

a)   b)

Figure 11.1: The gastrointestinal tract in the thoracic
and abdominal cavities, (a) anterior; (b) posterior
(note: the small intestine and part of the transverse colon have been removed)

Mouth
Food is chewed and mixed with saliva into a bolus. Salivary amylase begins the breakdown of starches and lingual lipase begins the breakdown of fats.

Stomach
Bolus is churned and mixed with gastric juice until it becomes chyme. Salivary amylase and lingual lipase continue to act for about one hour until they are denatured by HCL. Gastric lipase continues the breakdown of fat until it is also denatured by HCL. Pepsin begins the breakdown of proteins and in infants rennin begins the digestion of milk.

Esophagus
Bolus is moved to the stomach through peristalsis.

Liver
Produces bile which it secretes into the gall bladder. Also has numerous essential functions.

Pancreas
Secretes pancreatic juice into the duodenum of the small intestine. Pancreatic juice contains pancreatic amylase which continues the breakdown of carbohydrates, trypsin which continues the breakdown of proteins, pancreatic lipase which continues the breakdown of lipids and enzymes that digest nucleic acids.

Gall bladder
Stores bile and secretes it into the duodenum. Bile emulsifies fats.

Small intestine
Moves chyme by the actions of segmentation and peristalsis. Completes the digestion of most nutrients by a combination of pancreatic juice, bile, intestinal juice and brush border enzymes. Also absorbs most nutrients: the duodenum absorbs some micro-minerals; the jejunum absorbs water-soluble vitamins, amino acids, sugars, water and some minerals; and the ileum absorbs free fatty acids, cholesterol and fat-soluble vitamins.

Large intestine
Completes mechanical digestion through the movements of haustral churning, peristalsis and mass peristalsis. Bacteria complete all digestion and produce some B vitamins and vitamin K. Absorbs most of the water content of chyme and turns it into faeces, ready for elimination.

Organization of the Digestive System

Table 11.1: Organization of the digestive system

Organs and Structures of the GI Tract	Accessory Structures
• Mouth • Pharynx • Esophagus • Stomach • Small intestine, which is composed of the duodenum, jejunum, and ileum • Large intestine • Anus	• Teeth, tongue, and salivary glands, which are all located in the mouth • Liver • Gall bladder • Pancreas

Peritoneum

The peritoneum is a large serous membrane lining the abdominal cavity. Like the pleural and pericardial serous membranes, the peritoneum has two layers: the **parietal peritoneum** lining the walls of the abdominopelvic cavity and the **visceral peritoneum** (also called the serosa), which covers the organs of the digestive system.

Between the two layers of the peritoneum is a space called the **peritoneal cavity**. This is filled with serous fluid. The peritoneum is composed of large folds. These folds weave between the organs, binding them to each other and to the walls of the abdominal cavity.

These folds also contain many blood and lymphatic vessels and nerves, and include the:

- **Mesentery**, which binds the small intestine to the posterior abdominal wall
- **Mesocolon**, which binds the large intestine to the posterior abdominal wall
- **Falciform ligament**, which binds the liver to the anterior abdominal wall and diaphragm
- **Lesser omentum**, which links the stomach and duodenum with the liver
- **Greater omentum**, which covers the colon and small intestine.

Walls of the GI tract

The walls of the GI tract are composed of four layers of tissue. This basic arrangement of tissue does differ slightly with some of the organs, but in general it is composed of the **mucosa**, **submucosa**, **muscularis**, and **serosa**.

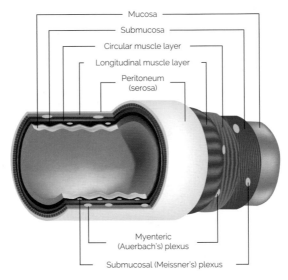

Mucosa
Submucosa
Circular muscle layer
Longitudinal muscle layer
Peritoneum (serosa)
Myenteric (Auerbach's) plexus
Submucosal (Meissner's) plexus

The wall of the gastrointestinal tract is composed of four layers: the inner mucosa lines the lumen of the tract, and layered on top of it is the submucosa, the muscularis and the serosa.

Figure 11.2: Structure of the wall of the GI tract

Table 11.2: Structure of the wall of the GI tract

Mucosa	The deepest layer of the GI tract. Its inner lining is a **mucous membrane**, which functions in protection, secretion, and absorption. This mucous membrane is attached to the **lamina propria**, which is a loose connective tissue layer that supports the blood and lymphatic vessels into which digested molecules are absorbed. The lamina propria also contains mucosa-associated lymphoid tissue (MALT), which protects the body against the many microbes and toxins that may have been ingested. The third layer of the mucosa is the **muscularis mucosa**, which is a layer of smooth muscle
Submucosa	Composed of areolar connective tissue and binds the mucosa to the muscularis layer. It contains many blood and lymphatic vessels and nerves and houses the **submucosal (Meissner's) plexus**. This is a network of nerves that serves the smooth muscle cells of the mucosa and forms part of the autonomic nervous system
Muscularis	A layer of muscle tissue. It includes some skeletal muscle tissue (especially in the mouth, pharynx, and upper portion of the esophagus, as these are the voluntary muscles of swallowing) and smooth muscle tissue. The smooth muscle tissue is composed of an inner sheet of circular fibers and an outer sheet of longitudinal fibers. These two sheets work together to physically break down and propel food along the GI tract. The muscularis also contains the myenteric plexus (plexus of Auerbach). This is a major nerve supply to the GI tract
Serosa	The most superficial layer of the wall of the GI tract, and it is a serous membrane. The esophageal portion of the serosa is called the **adventitia**, and it is composed of areolar connective tissue. The rest of the serosa is found below the diaphragm and is called the **visceral peritoneum**

Gastrointestinal Tract and Its Accessory Organs

Mouth (oral or buccal cavity)

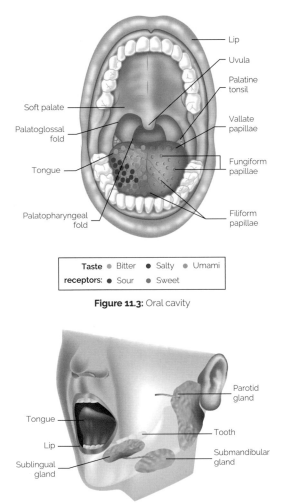

Figure 11.3: Oral cavity

Figure 11.4: Salivary glands

Table 11.3: Digestion in the mouth

Digestion/Enzyme	Description
Mechanical digestion The mechanical digestion of food begins in the mouth	
Chewing (mastication)	The tongue moves food, the teeth grind it, and saliva mixes with the food and begins to dissolve it; finally, it is reduced to a soft, flexible mass called a **bolus**
Swallowing (deglutition)	Swallowing is the mechanical process by which the bolus is moved from the mouth into the stomach; it involves the mouth, pharynx, and esophagus: • The tongue forces the bolus to the back of the oral cavity and into the oropharynx • Breathing is then temporarily interrupted as the respiratory passages are closed by the upward movement of the soft palate and the uvula • As this movement occurs, the larynx comes forward and upward over the tongue and the epiglottis moves backward and downward, thus closing off the space between the vocal folds (**rima glottidis**) • This enables the bolus to pass through the laryngopharynx and into the esophagus without entering the respiratory tract • The entire process takes only 1–2 seconds and then the respiratory passage reopens and breathing continues as normal
Chemical digestion The digestion of both carbohydrates and lipids begins in the mouth. However, because the food is in the mouth for such a short time, the digestion of these continues in the stomach	
Salivary amylase (in saliva)	Salivary amylase begins the breakdown of large carbohydrate molecules such as starch or polysaccharides into disaccharides and then monosaccharides
Lingual lipase (secreted by glands on the tongue)	Lingual lipase begins the breakdown of fats (lipids) from triglycerides into fatty acids and glycerol

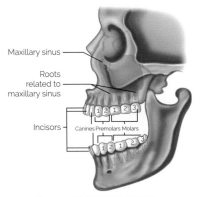

Maxillary sinus

Roots related to maxillary sinus

Incisors

Canines Premolars Molars

Permanent (secondary) dentition

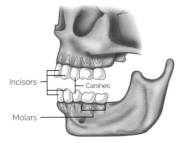

Incisors

Canines

Molars

a)   Deciduous (primary) dentition

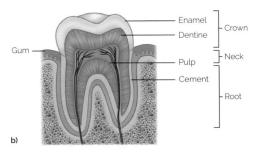

Enamel

Dentine

Crown

Gum

Pulp

Neck

Cement

Root

b)

Figure 11.5: (a) The teeth of an adult and a child; (b) cross-section of a tooth

Esophagus

The esophagus is a muscular tube approximately 10 in. (25 cm) long. It secretes mucus, which helps facilitate the movement of the bolus as it pushes it down toward the stomach through a process called **peristalsis**. No digestion takes place in the esophagus.

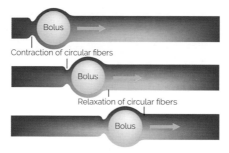

Figure 11.6: Movement of a bolus through the GI tract

Stomach

The stomach is a J-shaped organ that is continuous with the GI tract, and takes in food from the esophagus before breaking it down by means of enzymes and hydrochloric acid. It is divided into four regions: the cardia, fundus, body, and pylorus.

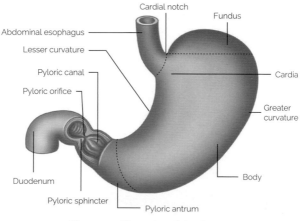

Figure 11.7: Stomach and duodenum

Table 11.4: Digestion in the stomach

Digestion/Enzyme	Description
Mechanical digestion In the stomach food is mixed, pummeled, and churned into a thin liquid called **chyme**	
Mixing waves	Every 15–25 seconds a mixing wave passes over the stomach, pushing its contents backward and forward and compressing and pummeling it into a liquid state
Chyme is "squirted" into the duodenum	The pyloric sphincter muscle between the stomach and duodenum is never completely closed, and as the stomach contents are pushed against the muscle, small amounts of chyme (approximately 3 ml) are forced or squirted out of the stomach and into the duodenum
	The remaining contents continue to be mixed and churned until the next wave forces a little more chyme into the duodenum
Chemical digestion In the presence of food, endocrine cells in the walls of the stomach secrete the hormone **gastrin**, which stimulates the production of gastric juice. Gastric juice is secreted by gastric glands and contains: • **Water**—liquefies the food • **Hydrochloric acid** (HCl)—kills microbes that have been ingested, partially denatures proteins, stimulates the secretion of hormones that promote the flow of bile and pancreatic juice, stops the action of salivary amylase and lingual lipase, and is needed for the conversion of pepsinogen into pepsin • **Intrinsic factor**—necessary for the absorption of vitamin B12 from the ileum • **Pepsinogen**—an enzyme precursor that is converted into pepsin in the acidic environment of gastric juice; pepsin is an enzyme that begins the breakdown of protein • **Gastric lipase**—an enzyme that acts on lipids, breaking triglycerides down into fatty acids and monoglycerides A 1–3 mm layer of mucus forms a protective barrier between the acidic gastric juice and the stomach wall, and also protects the epithelial cells of the wall from pepsin, which digests proteins	
Salivary amylase	The digestion of carbohydrates by salivary amylase continues in the stomach for about one hour before salivary amylase is denatured by hydrochloric acid

Table 11.4: (continued)

Digestion/Enzyme	Description
Lingual lipase	Similarly, the digestion of fats by lingual lipase continues for about one hour before the lingual lipase is denatured by hydrochloric acid
Gastric lipase	Continues the breakdown of fats; however, gastric lipase is soon denatured by the acidity of gastric juice
Pepsin	Pepsin begins the breakdown of proteins
Rennin	Rennin is an enzyme found only in the stomachs of infants; it begins the digestion of milk by converting the protein caseinogen into casein

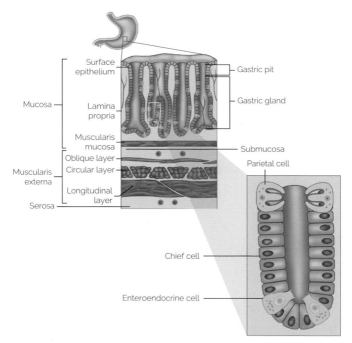

Figure 11.8: Stomach wall

Liver

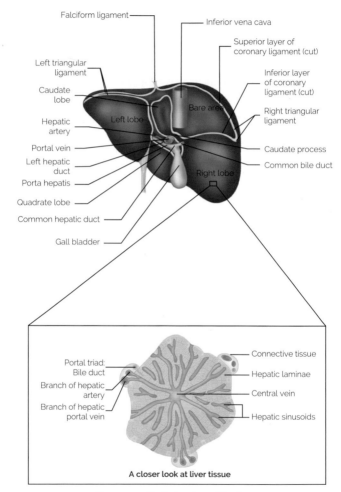

Falciform ligament

Inferior vena cava

Superior layer of coronary ligament (cut)

Left triangular ligament

Inferior layer of coronary ligament (cut)

Caudate lobe

Bare area

Right triangular ligament

Hepatic artery

Left lobe

Portal vein

Caudate process

Left hepatic duct

Common bile duct

Porta hepatis

Right lobe

Quadrate lobe

Common hepatic duct

Gall bladder

Connective tissue

Portal triad: Bile duct

Hepatic laminae

Branch of hepatic artery

Central vein

Branch of hepatic portal vein

Hepatic sinusoids

A closer look at liver tissue

Figure 11.9: Posterior view of the liver

The liver is composed of two lobes separated by the **falciform ligament**. Each lobe is composed of smaller functional units called **lobules**, and each lobule is made up of specialized epithelial cells called **hepatocytes**. These cells are arranged around a central vein.

The liver does not contain capillaries. Instead, it has spaces through which blood passes. These spaces are called **sinusoids**, and they are partly lined by phagocytes, which destroy microbes and potentially harmful foreign matter.

The liver also has a unique double blood supply. It receives oxygenated blood via the **hepatic artery**. It also receives deoxygenated blood from the GI tract via the **hepatic portal vein**. This blood contains newly absorbed nutrients as well as drugs, toxins, or microbes that may have been absorbed from the GI tract. This blood needs to be "cleaned," or "made safe," by the liver before it can be circulated to the rest of the body.

Table 11.5: Functions of the liver

Carbohydrate metabolism	The liver maintains normal blood glucose levels
Lipid metabolism	The liver stores fats and, when necessary, converts them into a form that can be used by the tissues. It also breaks down fatty acids; synthesizes lipoproteins, which are necessary for transporting fatty acids, triglycerides, and cholesterol; and synthesizes cholesterol
Protein metabolism	The liver synthesizes all the major plasma proteins, and the processes of transamination and deamination take place in the liver. The liver also converts ammonia into urea for excretion
Detoxification	The liver removes and excretes alcohol and some drugs and also chemically alters and excretes some hormones
Storage of nutrients	The liver stores glycogen; vitamins A, B12, D, E, and K; and the minerals iron and copper
Phagocytosis	Some liver cells phagocytize worn-out red blood cells, white blood cells, and some bacteria
Activation of vitamin D	Together with the skin and kidneys, the liver participates in activating vitamin D
Production of bile	The liver produces bile and secretes it into the gall bladder for storage. Bile contains water, bile acids, bile salts, cholesterol, phospholipids, bile pigments, and some ions

Gall bladder

The gall bladder is a pear-shaped, green sac that is located behind the liver and attached to it by connective tissue (see figure 11.9). The gall bladder receives bile from the liver and concentrates and stores it. It then releases bile into the duodenum via the common bile duct.

Small intestine

The small intestine is made up of three segments:

- **The duodenum:** This is the first segment of the small intestine and is also the shortest. It is approximately 10 in. (25 cm) long and it receives food from the stomach, bile from the gall bladder, and pancreatic juice from the pancreas.
- **The jejunum:** This is approximately 40 in. (1 m) long and lies between the duodenum and the ileum.
- **The ileum:** This is the longest segment of the small intestine at approximately 80 in. (2 m) long. It receives food from the jejunum and passes it into the large intestine via the **ileocecal valve**.

The structure of the wall of the small intestine is similar to that of the rest of the GI tract in that it has the same basic four layers of tissue (mucosa, submucosa, muscularis, and serosa). However, because most digestion and absorption takes place in the small intestine, it is uniquely structured to ensure a large surface area. About 90% of all absorption takes place in the small intestine.

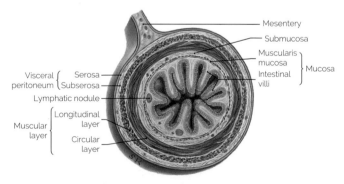

Figure 11.10: Cross-section through the small intestine

Table 11.6: Structure of the walls of the small intestine

Structure	Description	Function
Plicae circulares (circular folds)	Mucosa has ridges that are approximately 10 mm high	Increase surface area and cause chyme to spiral
Villi	Mucosa forms fingerlike projections called villi (singular = **villus**). Each villus contains an arteriole, venule, capillary network, and lacteal	Increase surface area and enable absorbed nutrients to enter bloodstream and lymphatic system
Microvilli	Every villus contains absorptive cells. Each absorptive cell has many tiny, membrane-covered projections called microvilli (singular = microvillus)	Increase surface area for absorption
Brush border	Microvilli form the brush border, which contains enzymes that have been secreted into the plasma membranes of the microvilli instead of into the lumen of the small intestine	Brush-border enzymes continue the breakdown of disaccharides into monosaccharides, complete the digestion of proteins, and help digest nucleotides
Intestinal glands (crypts of Lieberkühn)	Mucosa contains cavities called intestinal glands. They secrete intestinal juice and contain cells, including Paneth cells, enteroendocrine cells, and goblet cells	Intestinal juice helps bring nutrient particles into contact with microvilli; Paneth cells secrete a bactericidal enzyme called lysozyme; enteroendocrine cells secrete hormones; and goblet cells secrete mucus
Peyer's patches	Located in the lamina propria of the mucosa	Function as mucus-associated lymphoid tissues
Duodenal glands (Brunner's glands)	Located in the submucosa of the duodenum	Secrete an alkaline mucus that helps neutralize gastric acid in the chyme

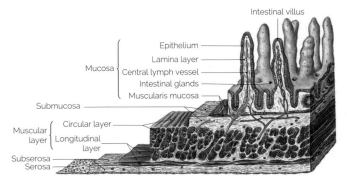

Figure 11.11: Layers of the wall of the small intestine

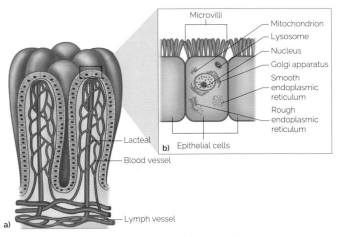

Figure 11.12: Arrangement of (a) villi and (b) microvilli

Table 11.7: Digestion in the small intestine

Digestion/Enzyme	Description
Mechanical digestion In the small intestine, the thin liquid chyme is mixed with digestive juices and brought into contact with the villi and microvilli, where it can be further digested and absorbed. It is moved in and through the small intestine by two movements: segmentation and peristalsis	
Segmentation	• Segmentation is the main movement in the small intestine • It involves localized contractions that move food back and forth and is a similar movement to squeezing opposite ends of a tube of toothpaste so that the paste is moved back and forth
Peristalsis	• Peristalsis in the small intestine is a weak movement compared to that in the esophagus or stomach • It slowly propels chyme forward toward the large intestine
Chemical digestion Chyme entering the small intestine contains partially digested nutrients, which are then broken down further by a combination of pancreatic juice, bile, intestinal juice, and brush-border enzymes. Digestion and absorption of most nutrients is usually completed in the small intestine	
Carbohydrate digestion	
Pancreatic amylase	Present in pancreatic juice and completes the breakdown of starches and glycogen; however, it does not digest cellulose, which passes into the large intestine as fiber
α-dextrinase	This brush-border enzyme also acts on starches and breaks them down into glucose units
Maltase	This brush-border enzyme breaks down maltose into glucose
Sucrase	This brush-border enzyme breaks down sucrose into glucose and fructose
Lactase	This brush-border enzyme breaks down lactose into glucose and galactose

Table 11.7: (continued)

Digestion/Enzyme	Description
Protein digestion	
Trypsin, chymotrypsin, carboxypeptidase, elastase	These enzymes, present in pancreatic juice, break down proteins into peptides
Peptidases (aminopeptidase and dipeptidase)	These brush-border enzymes complete the breakdown of proteins into amino acids
Lipid digestion	
Bile salts	Emulsify lipids; i.e., they break down large globules of triglycerides into smaller droplets—this exposes a greater surface area to the enzyme pancreatic lipase
Pancreatic lipase	Present in pancreatic juice and breaks down triglycerides into fatty acids and monoglycerides
Nucleic acid digestion	
Ribonuclease	Present in pancreatic juice and breaks down RNA into nucleotides
Deoxyribonuclease	Present in pancreatic juice and breaks down DNA into nucleotides
Nucleosidases, phosphatases	These brush-border enzymes then break nucleotides down into pentoses, phosphates, and nitrogenous bases

Table 11.8: Absorption in the small intestine

Region of Small Intestine	Nutrients Absorbed
Carbohydrates can only be absorbed as monosaccharides (glucose, fructose, galactose); proteins as amino acids, dipeptides, and tripeptides; and lipids as fatty acids, glycerol, and monoglycerides	
Duodenum	Microminerals
Jejunum	Water-soluble vitamins, amino acids, sugars, water, and some minerals
Ileum	Free fatty acids, cholesterol, fat-soluble vitamins, and bile

Pancreas

The pancreas is a long, thin gland lying behind the stomach and connected to the duodenum by two ducts, the larger **pancreatic duct** (**duct of Wirsung**), which joins the common bile duct from the liver, and the smaller **accessory duct** (**duct of Santorini**).

Exocrine cells secrete **pancreatic juice,** which is a clear liquid composed of mostly water, some salts, sodium bicarbonate, and some enzymes. It is slightly alkaline and so buffers the acidic chyme coming from the stomach, stops the action of pepsin, and creates the correct pH in the small intestine in which the enzymes here can work.

Endocrine cells are also found in clusters called **pancreatic islets** (islets of Langerhans). These secrete hormones such as glucagon and insulin.

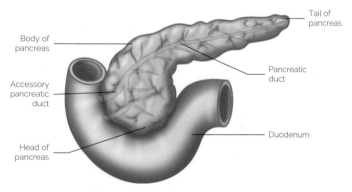

Body of pancreas

Accessory pancreatic duct

Head of pancreas

Tail of pancreas

Pancreatic duct

Duodenum

Figure 11.13: Duodenum and pancreas

Large intestine

The structure of the wall of the large intestine is similar to that of the rest of the GI tract in that it contains all four tissue layers (mucosa, submucosa, muscularis, and serosa). However, the mucosa of the colon contains absorptive cells that absorb water, and goblet cells that secrete mucus to lubricate the contents of the large intestine.

The muscularis also has a unique arrangement of muscle fibers called **taeniae coli**. These are always slightly contracted to gather the colon into pouches called **haustra**.

The absorption of water and electrolytes takes place in the ascending and transverse colon, and liquid chyme becomes a solid or semi-solid mass called feces. Feces contain water, inorganic salts, sloughed-off epithelial cells from the mucosa of the GI tract, bacteria, products of bacterial decomposition, and undigested foods. Feces are eliminated from the body by a process called **defecation**.

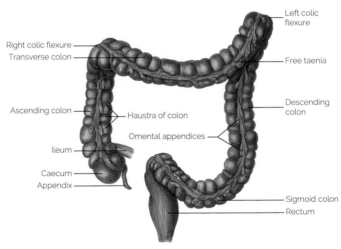

Figure 11.14: Large intestine

Table 11.9: Regions of the large intestine

Cecum	A 2¼ in. (6 cm) pouch that receives food from the small intestine via the ileocecal valve (ileocecal sphincter). Attached to the cecum is the vermiform appendix
Colon	Forms most of the large intestine and is a long tube made up of the ascending colon, transverse colon, descending colon, and sigmoid colon. The sigmoid colon is the last part of the colon, joining the rectum at the level of the third sacral vertebra
Rectum	Approximately 8 in. (20 cm) long and lies in front of the sacrum and coccyx
Anal canal	The last inch (2–3 cm) of the rectum. Its external opening is called the anus and it is guarded by both internal and external sphincter muscles. These are usually closed except during defecation

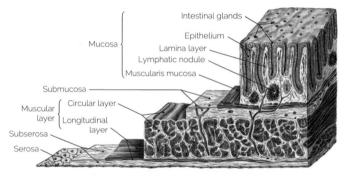

Figure 11.15: Layers of the wall of the colon

Table 11.10: Digestion in the large intestine

Digestion/Enzyme	Description
Mechanical digestion Mechanical digestion in the large intestine is very slow and consists of three movements: haustral churning, peristalsis, and mass peristalsis	
Haustral churning	The haustra, or pouches of the colon, are relaxed and distended until they are filled with matter, then they contract and squeeze their contents into the next haustrum. Thus, the contents of the large intestine are slowly moved from haustrum to haustrum
Peristalsis	Peristalsis in the large intestine is a very weak, slow movement

Table 11.10: (*continued*)

Digestion/Enzyme	Description
Mass peristalsis	Three to four times a day, usually during or immediately after a meal, a very strong peristaltic wave begins in the middle of the transverse colon; it quickly travels across the rest of the colon, driving the colonic contents into the rectum
Chemical digestion As mentioned earlier, most digestion and absorption of nutrients is completed in the small intestine. However, small amounts of undigested nutrients do pass into the large intestine, and these are digested by bacteria living in the lumen	
Bacteria	Bacteria in the lumen prepare the chyme for elimination by: • Fermenting any remaining carbohydrates • Converting any remaining proteins into amino acids and breaking down amino acids into simpler substances • Decomposing bilirubin into simpler substances The bacteria also produce some B vitamins as well as vitamin K, which are absorbed in the colon

Digestion of carbohydrates

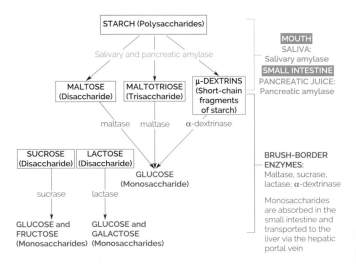

Digestion of proteins

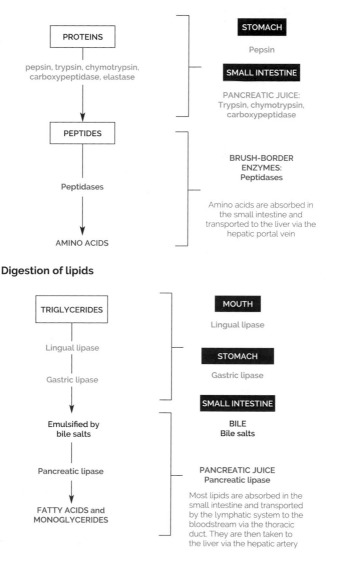

PROTEINS

pepsin, trypsin, chymotrypsin,
carboxypeptidase, elastase

PEPTIDES

Peptidases

AMINO ACIDS

STOMACH

Pepsin

SMALL INTESTINE

PANCREATIC JUICE:
Trypsin, chymotrypsin,
carboxypeptidase

BRUSH-BORDER
ENZYMES:
Peptidases

Amino acids are absorbed in
the small intestine and
transported to the liver via the
hepatic portal vein

Digestion of lipids

TRIGLYCERIDES

Lingual lipase

Gastric lipase

Emulsified by
bile salts

Pancreatic lipase

FATTY ACIDS and
MONOGLYCERIDES

MOUTH

Lingual lipase

STOMACH

Gastric lipase

SMALL INTESTINE

BILE
Bile salts

PANCREATIC JUICE
Pancreatic lipase

Most lipids are absorbed in the
small intestine and transported
by the lymphatic system to the
bloodstream via the thoracic
duct. They are then taken to
the liver via the hepatic artery

Table 11.11: Chemical digestion of carbohydrates, proteins, and lipids

Nutrient	Enzymes	From	To
Carbohydrates Begins in mouth and completed in small intestine	Amylases	Starches, polysaccharides, and disaccharides	Monosaccharides
Proteins Begins in stomach and completed in small intestine	Proteases	Peptones and polypeptides	Amino acids
Lipids Minimal digestion in mouth and stomach. Most occurs in small intestine	Lipases	Triglycerides	Fatty acids and glycerol

Theory in practice

Embedded in the lining of the gut is the enteric nervous system. It contains as many nerve cells as there are in the spinal cord. There are more immune cells in the wall of the gut than in the bloodstream and bone marrow combined; it contains a huge number of endocrine cells, which release up to 20 different hormones; and it is the largest storage facility of serotonin in the body.

This system controls all digestive processes and is itself affected by the central nervous system. This gut-brain connection explains why you get "butterflies" in your stomach when you are nervous, or nausea when you are anxious or scared. In addition, simply thinking of, or smelling, food can trigger the release of digestive juices. It is no surprise, then, that when under emotional or mental stress we can develop functional bowel disorders such as pain, bloating, diarrhea, or constipation, even when there is no structural cause.

In addition to the enteric nervous system, according to neuroscientist and gastroenterologist Emeran Mayer "there are 100,000 times more microbes in your gut alone as there are people on earth." These microbes not only assist in digesting food but also help regulate metabolism, detoxify chemicals, and train the immune system. No wonder what you eat affects your health and mood so much!

CHAPTER 12

The Urinary System

The kidneys control the composition, volume, and pressure of blood; help regulate blood pH; help synthesize calcitriol, which is the active form of vitamin D; secrete erythropoietin, which stimulates the production of red blood cells; and, during periods of starvation, can synthesize new glucose molecules in a process called **gluconeogenesis**. Other organs of the urinary system (the ureters, bladder, and urethra) transport, store, and excrete urine.

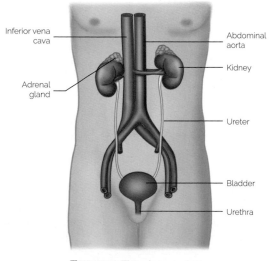

Inferior vena cava

Abdominal aorta

Kidney

Adrenal gland

Ureter

Bladder

Urethra

Figure 12.1: The urinary system

Kidneys

The kidneys are uniquely structured to filter blood and produce urine. Together, the renal cortex and renal pyramids form the functional part of the kidney. They are composed of approximately one million microscopic structures called **nephrons**. These are the functional units of the kidney and are where urine is formed. Nephrons have three functions (filtration, secretion, and reabsorption), and it is through these functions that the composition and volume of blood are regulated and urine is produced.

The reabsorption of water and electrolytes by the kidneys is regulated by hormones. The remaining areas of the kidneys function as passageways and storage areas. Blood is brought to the kidneys by the right and left renal arteries and drained by the renal veins.

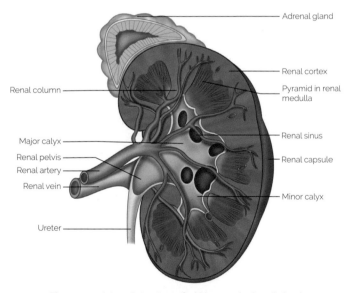

Figure 12.2: Internal structure of a kidney and adrenal gland

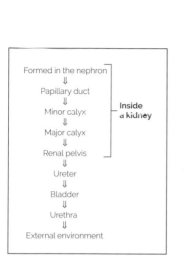

Flow of urine

Flow of blood through the kidneys

Table 12.1: Analysis of normal urine

Characteristic	Description
Volume	1¾–3½ pints (1–2 liters) every 24 hours
Color	Clear to pale yellow
	Color due to pigments can vary
Odor	Initially slightly aromatic but becomes ammonia-like upon standing
	Odor can vary
pH	Varies considerably according to diet, and ranges between 4.6 and 8.0
Solute	Description
Urea	The main product of protein metabolism
Creatinine	Product of muscle activity
Uric acid	Product of nucleic acid metabolism
Urobilinogen	Bile pigment derived from the breakdown of hemoglobin
Inorganic ions	These vary with one's diet

Nephron

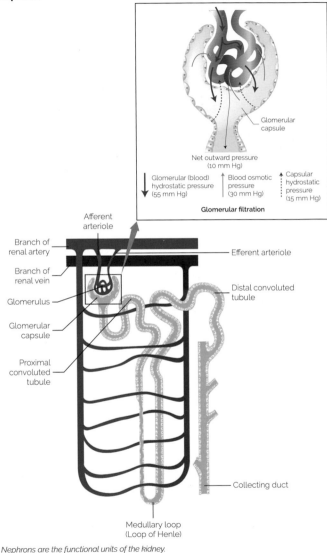

Nephrons are the functional units of the kidney.

Figure 12.3: A nephron and associated blood vessels

Table 12.2: Structure of a nephron

Structure	Description	Function
Renal corpuscle	Composed of a knotted network of capillaries called the glomerulus and a cuplike structure surrounding the glomerulus called the glomerular (Bowman's) capsule, which is the closed end of the renal tubule	Filters blood plasma. Blood flows through the capillaries in the glomerulus and water and most solutes filter from the blood into the glomerular capsule across a membrane called the filtration membrane. However, large plasma proteins and formed elements such as red and white blood cells are too big to filter through the walls of the glomerulus and so remain in the blood
Renal tubule	Composed of three sections: The proximal convoluted tubule (PCT), where the reabsorption of most substances takes place; the loop of Henle (nephron loop), where water and salts are reabsorbed; and the distal convoluted tubule (DCT), where "fine-tuning" of the filtrate occurs	Functions in secretion and reabsorption

Table 12.3: Hormonal regulation of fluid and electrolyte balance

Endocrine Gland	Hormone	Regulation
Parathyroid glands	Parathyroid hormone (parathormone)	Increases blood calcium and magnesium levels and decreases blood phosphate levels
Thyroid gland	Calcitonin	Lowers blood calcium levels
Adrenal cortex	Aldosterone	Increases blood levels of sodium and water and decreases blood levels of potassium. Because sodium is responsible for the osmotic flow of water, aldosterone helps regulate the amount of water in the blood. The release of aldosterone is stimulated by decreased sodium levels or increased potassium levels in the extracellular fluid; and by the renin-angiotensin mechanism
Posterior pituitary gland	Antidiuretic hormone (vasopressin)	Decreases the secretion of urine by causing the collecting ducts of the kidneys to reabsorb more water. When this additional water is returned to the bloodstream, it increases the volume, and therefore pressure, of the blood

Ureters

The ureters are two 10–12 in. (25–30 cm) tubes that carry urine from the kidneys to the bladder. Each ureter drains the renal pelvis of a kidney and inserts into the posterior aspect of the bladder.

The walls of the ureters are composed of three layers of tissue: the **adventitia** (outer layer), the **muscle** (intermediate layer), and **mucosa** (inner layer). They do not have a physical valve that prevents the backflow of urine from the bladder into the ureters. However, when the bladder is full, pressure from within it compresses the ureter openings. This, coupled with the force of gravity, prevents any urinary backflow.

Urinary bladder

The urinary bladder, which is more commonly referred to as the bladder, is a freely movable, hollow muscular organ that is held in its place by folds of the peritoneum. The bladder acts as a reservoir, storing urine until it is excreted out of the body. On the floor of the bladder is a triangular area called the **trigone**. The openings to the ureters are located in the posterior two corners of the trigone and the urethral opening, the **internal urethral orifice**, is located in the anterior corner.

Similar to the walls of the ureters, the walls of the bladder consist of three layers of tissue: the **adventitia** (outer layer), the **detrusor** muscle, including the outer longitudinal, inner longitudinal, and middle circular muscles (intermediate layer), and the **mucosa** (inner layer).

The bladder also has two sphincter muscles, an internal urethral sphincter and an external urethral sphincter, which regulate **micturition** (emptying or voiding of the bladder). When the bladder is approximately half full, stretch receptors in the wall of the bladder send messages to the spinal cord. From here a reflex arc returns to the internal urethral sphincter causing it to relax and empty the bladder. This is known as the **micturition reflex**. Although emptying the bladder is a reflex action, the external urethral sphincter is controlled by striated muscle. Thus, emptying the bladder becomes a learned, voluntary action.

Urethra
At the end of the urinary system is a passageway that functions in discharging urine from the body. This passageway is the urethra, and it is a small tube leading from the internal urethral orifice in the bladder to the external environment.

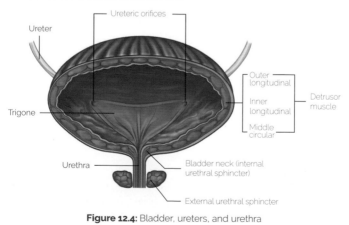

Figure 12.4: Bladder, ureters, and urethra

Theory in practice

Water is vital and without it we would not survive. Every cell in our body contains water, the size and shape of every cell are maintained by water; it is the solvent in all body fluids and the medium for all biochemical processes. It is so vital that more than half our body weight comes from water! It circulates in blood plasma, and as it circulates it gathers toxins, wastes, and any other substances that are not used by the cells. Blood, therefore, needs to be continually cleaned and its water content regulated, otherwise tissue cells would "drown" in their own water and blood would become a river of circulating toxins and waste products. This vital role of cleaning and regulating blood is carried out by the two kidneys.

CHAPTER 13

The Reproductive System

The human reproductive system functions in reproducing life and continuing the species. Reproductive cells are called **gametes** and are produced in gonads (testes or ovaries) of men and women through a type of cell division called **meiosis** (see Chapter 2).

Meiosis results in four daughter cells that each have only one set of chromosomes (23 chromosomes). These are called **haploid cells**. A male gamete formed in the male reproductive system then enters the female reproductive system through sexual intercourse. The male and female gametes unite and fuse in a process called **fertilization**. This produces a **zygote**, which is a new cell that now contains two sets of chromosomes (46 chromosomes)—one set from the mother and one from the father. The zygote then begins to divide by mitosis and develops into a new organism.

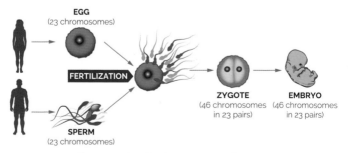

Figure 13.1: Fertilization and zygote formation

Male Reproductive System

The male reproductive system consists of the testes, which are the gonads where spermatozoa are formed through the process of spermatogenesis; a system of ducts for transporting and storing sperm; and accessory organs that produce supporting substances.

The testes produce male sex hormones called **androgens**.
The principal androgen is testosterone. It stimulates the development of masculine secondary sex characteristics; promotes growth and maturation of the male reproductive system and sperm production; promotes male sexual behavior and libido; and stimulates anabolism resulting in heavier muscle and bone mass.

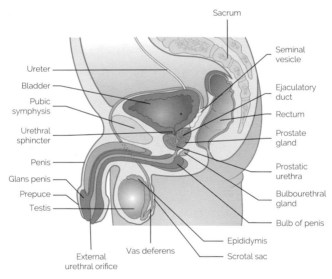

Figure 13.2: Overview of the male urinary and genital systems (mid-sagittal view)

Table 13.1: Male reproductive system

Structure	Description	Function
Scrotum	A sac of loose skin and superficial fascia in which lie the testes. The scrotum hangs from the root of the penis and is divided internally into two sacs. Each sac contains one testis	Houses the testes and maintains a temperature approximately 3°C cooler than normal body temperature
Testes	Pair of oval glands located in scrotum. Each testis is enclosed by a fibrous tunica albuginea and covered by the tunica vaginalis. Extensions of this divide each testis into internal lobules containing seminiferous tubules where spermatogenesis occurs and sperm are formed. Sustentacular cells are located in the seminiferous tubules and interstitial endocrinocytes (Leydig cells) are located between them	Spermatogenesis occurs in the testes. Sustentacular cells help regulate effects of testosterone and follicle-stimulating hormone (FSH); protect, support, and nourish sperm; and produce fluid in which sperm are transported. Leydig cells secrete testosterone
Epididymis	A comma-shaped organ lying along the posterior border of each testis and composed of a series of coiled ducts that empty into a single tube called the ductus epididymis	Site of sperm maturation. Stores sperm until they are fully mature and then helps propel them via peristaltic contractions
Vas deferens (ductus deferens, seminal duct)	Long duct, which runs from the epididymis into the pelvic cavity	Transports sperm via peristaltic contractions from epididymis to urethra
Spermatic cord	A supporting structure consisting of blood vessels, lymphatic vessels, nerves, and muscles. It runs alongside the vas deferens	Supports the vas deferens
Urethra	Terminal duct of both the reproductive and urinary systems and it is made up of three sections: prostatic urethra, membranous urethra and spongy (penile) urethra	Transports semen and urine out of the body

Table 13.1: (*continued*)

Structure	Description	Function
Seminal vesicles	Paired structures located at the base of the bladder	Secrete a viscous alkaline fluid that contains fructose, prostaglandins, and clotting proteins
Prostate gland	Gland that surrounds the prostatic urethra	Secretes a fluid that contributes to sperm mobility and viability
Bulbourethral (Cowper's) glands	Paired structures located on either side of the membranous urethra	Secrete a lubricating mucus and alkaline substance that neutralizes the acidity of urine and lubricates the end of the penis
Penis	Composed of erectile tissue permeated by blood sinuses. When stimulated, arteries dilate and blood enters the sinuses, which then expand and compress veins that normally drain the penis. Blood in penis is trapped and penis becomes erect. Made up of three regions: root, shaft, and glans penis	Excretes urine and ejaculates semen. During ejaculation, the sphincter muscle at the base of the urinary bladder closes to prevent any urine passing into the urethra
Perineum	Diamond-shaped area that contains the external genitals and the anus	

Female Reproductive System

Female gonads are ovaries, which produce ova through the process of oogenesis. The female reproductive system functions in producing ova, receiving spermatozoa, being the site of fertilization, and housing the fertilized zygote so that it can grow and develop into a fetus. This is called pregnancy. The female reproductive system also functions in delivering the fetus into the external world through the process of childbirth (labor).

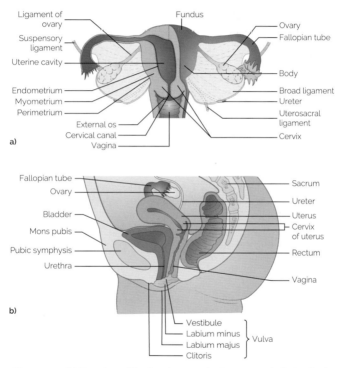

Figure 13.3: (a) Overview of the female reproductive organs (anterior view); (b) overview of the female urinary and genital systems (mid-sagittal view)

Table 13.2: Female reproductive system

Structure	Description	Function
Ovaries	Paired almond-shaped organs located in the superior portion of the pelvic cavity on either side of the uterus. They are held in place by a series of ligaments and a fold of the peritoneum called the broad ligament. Inside the ovaries are small saclike structures called ovarian follicles. Each follicle houses an immature ovum, called an oocyte. One follicle becomes the dominant follicle and is called the mature Graafian follicle. At ovulation this follicle expels the ovum into the uterine cavity. The now-empty follicle develops into the corpus luteum, a glandular structure that secretes female hormones. The corpus luteum finally degenerates into the corpus albicans	Gonads where ova are produced through oogenesis. Also function in ovulation. Ovulation is the process by which a mature Graafian follicle releases an ovum. The released ovum then travels down the Fallopian tube toward the uterus
	During fetal development	Germ cells differentiate into millions of immature eggs called primary oocytes. At birth, a woman has all the oocytes she will ever have
	From puberty onward	The release of gonadotropic hormones at puberty stimulates meiosis of a primary oocyte. This occurs in one oocyte, in one follicle, each month after puberty. The primary oocyte develops into a secondary oocyte, or ovum, and the follicle in which this development occurs is the mature Graafian follicle

Table 13.2: (continued)

Structure	Description	Function
Fallopian (uterine) tubes	Two thin tubes running from the ovaries to the uterus. Composed of an outer serous membrane, intermediate muscularis, and internal mucosa lined with cilia that help move a fertilized ovum	Site of fertilization, when sperm and ovum unite and fuse to form a zygote. Transport the zygote to the uterus. It takes approximately seven days for the zygote to travel down a tube and into the uterus
Uterus (womb)	A muscular sac located between the bladder and rectum. Approximately the size and shape of an inverted pear and divided into the fundus, body, and cervix. The cervix is lined with mucus-secreting cells that produce a mixture of water, proteins, lipids, enzymes, and inorganic salts. This is called cervical mucus, and approximately 20–60 ml is produced each day. The walls of the uterus are composed of three layers of tissue: an outer perimetrium (serosa), an intermediate myometrium, and an inner endometrium. The endometrium is a highly vascularized layer of tissue that forms the lining of the uterus and is divided into the: – **Stratum functionalis:** Functional layer of the endometrium and the layer closest to the uterine cavity. It is shed during menstruation – **Stratum basalis:** Permanent base layer of the endometrium. Produces a new stratum functionalis after each menstruation	Before fertilization, the uterus acts as a pathway through which sperm travel into the Fallopian tubes, where they attempt to fertilize an ovum. If fertilization is successful, then the uterus becomes the site of implantation of a zygote and houses the developing fetus throughout pregnancy. It then contracts forcefully during labor to expel the fetus. If, however, fertilization is not successful, the uterus becomes the site of menstruation. This is the process by which the lining of the uterus, the stratum functionalis, is shed and discarded

Table 13.2: (continued)

Structure	Description	Function
Vagina	Muscular tube located between the bladder and rectum and attached to the uterus. Its walls are composed of an outer adventitia, an intermediate muscularis, and an inner mucosa. The mucosa is continuous with that of the uterus and it lies in a series of transverse folds called **rugae**. Secretes an acidic mucus, which retards microbial growth but which also harms sperm. The opening of the vagina to the external environment is called the **vaginal orifice**. It is protected by a thin mucous membrane called the **hymen**, which partially occludes it	The vagina acts as a passageway for blood during menstruation, semen during sexual intercourse, and the fetus during childbirth
Vulva (external female genitalia)	Folds surrounding the opening to the vagina; consist of the mons pubis, labia majora, labia minora, clitoris, and vestibule. The vestibule houses several ducts including those from the mucus-secreting paraurethral (Skene's) glands and the greater vestibular (Bartholin's) glands	
Perineum	Diamond-shaped area that contains the external genitals and the anus	

Mammary glands

The mammary glands, or breasts, are two modified sudoriferous glands located over the pectoralis major muscles and attached to them by a layer of dense irregular connective tissue. Internally, a breast is supported by strands of connective tissue called **suspensory (Cooper's) ligaments** and is composed of compartments, or lobes, separated by adipose tissue.

Inside each lobe are smaller **lobules,** which contain milk-secreting **alveolar glands. These** secrete milk into secondary tubules, which drain into the **mammary ducts**. Milk is then stored temporarily and excreted externally through **lactiferous ducts**, via the nipple. The mammary glands synthesize and secrete milk through the process of **lactation,** which is stimulated by the hormone **prolactin** (with smaller contributions from progesterone and estrogens). The ejection of milk is stimulated by the hormone **oxytocin,** whose release is stimulated by the suckling action of the baby on the breast.

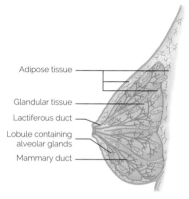

Adipose tissue

Glandular tissue

Lactiferous duct

Lobule containing alveolar glands

Mammary duct

Figure 13.4: Mammary gland

Table 13.3: Summary of the female reproductive hormones

Estrogens and **progesterone** are the principal female sex hormones and are produced mainly by the ovaries, although small amounts are also produced by the adrenal cortex, testes, and placenta. Women also produce androgens, which contribute to their libido. However, they are produced in far smaller quantities than in men and are produced by the adrenal cortex and not by the female gonads	
Estrogens	Necessary for reproductive and sexual development
Progesterone	Prepares the body for pregnancy
Testosterone	Increases bone and muscle strength as well as sexual desire
Follicle-stimulating hormone (FSH)	Stimulates ovarian follicles to grow
Luteinizing hormone (LH)	Stimulates rupture of the mature Graafian follicle and development of the corpus luteum
Inhibin	Inhibits secretion of FSH
Relaxin	Helps prepare a pregnant woman's body for childbirth

Female reproductive cycle

Table 13.4: Female reproductive cycle

Hormonal Regulation	Ovarian Cycle	Uterine (Menstrual) Cycle
Menstrual phase (menstruation, menses) The word *menses* means "month," and menstruation marks the beginning of a woman's monthly cycle. Menstruation normally lasts approximately 5 days and the first day of menstruation is termed day 1 of a woman's cycle Before menstruation begins, a woman's uterus, especially the stratum functionalis of the endometrium, is prepared to receive a fertilized ovum. If it does not receive a fertilized ovum, levels of estrogens and progesterone decline and the stratum functionalis dies and is discharged from the body via menstrual flow		
• Declining levels of estrogens and progesterone cause uterine arteries to constrict and endometrial cells to become deficient in blood • These cells eventually die and the entire stratum functionalis of the endometrium is sloughed off	Approximately 20 small follicles, some in each ovary, begin to enlarge	Menstrual flow (consisting of blood, tissue fluid, mucus, and epithelial cells derived from the endometrium) is discharged from the vagina
Preovulatory phase The preovulatory phase is the time between menstruation and ovulation. In a 28-day cycle it is usually 6–13 days in length. In this phase, a mature Graafian follicle forms and the endometrium proliferates		
• Follicle-stimulating hormone (FSH) from the anterior pituitary gland stimulates follicles to grow	• The follicles continue to develop and around day 6, one follicle in one ovary outgrows the other follicles	• Estrogen secreted by the follicles stimulates the repair of the endometrium, which now thickens and proliferates

Table 13.4: (continued)

Hormonal Regulation	Ovarian Cycle	Uterine (Menstrual) Cycle
• These growing follicles then secrete higher levels of estrogens and inhibin, which in turn decrease FSH secretion	• This becomes the dominant follicle, which is now called the mature Graafian follicle (vesicular follicle) and continues to enlarge until ovulation • The other follicles begin to degenerate • The menstrual and preovulatory phases are called the **follicular phase** of the ovarian cycle	• This phase is also called the **proliferative phase** of the uterine cycle

Ovulation

Ovulation usually occurs around day 14 of a 28-day cycle and involves the rupture of the mature Graafian follicle and the release of an ovum into the pelvic cavity. It is during this phase that a woman can now become pregnant

• High levels of estrogens stimulate the hypothalamus to release gonadotropin-releasing hormone (GnRH), which stimulates the anterior pituitary gland to release FSH and luteinizing hormone (LH) • Note that progesterone levels are now low	• LH stimulates the rupture of the mature Graafian follicle and the release of the ovum into the pelvic cavity • The follicle then develops into the corpus luteum, which, under the influence of LH, secretes progesterone, estrogens, relaxin, and inhibin	The prepared endometrium awaits the arrival of a fertilized ovum

Table 13.4: (continued)

Hormonal Regulation	Ovarian Cycle	Uterine (Menstrual) Cycle

Postovulatory phase

The 14 days after ovulation form the postovulatory phase. This is a "waiting time" in which the endometrium awaits the arrival of a fertilized ovum. It is now thickened, highly vascularized, and secreting tissue fluid and glycogen. Thus, this phase is also called the **secretory phase** of the uterine cycle, or the **luteal phase** of the ovarian cycle.

If fertilization has occurred, the ovum takes approximately a week to arrive at the endometrium, where it becomes embedded and develops into a fetus. At this stage a woman is now pregnant.

If fertilization has not occurred, menstruation and the reproductive cycle begin again

- The corpus luteum survives approximately two weeks and secretes increasing amounts of progesterone and some estrogens
- If fertilization has occurred, the corpus luteum (and the hormones it secretes) remains longer than two weeks and is maintained by human chorionic gonadotropin (hCG), a hormone produced by the embryo approximately 8–12 days after fertilization
- Once the embryo is implanted in the endometrium, the hormones of pregnancy come into effect
- If fertilization has not occurred, the corpus luteum degenerates and the lack of progesterone and estrogens causes menstruation

It is easy to become confused when learning about the female reproductive cycle, so it helps to remember that this cycle actually involves two smaller cycles: the ovarian cycle and the uterine cycle:

	Before Ovulation	Ovulation	After Ovulation
Uterine cycle	Menstrual and proliferative phases		Secretory phase
Ovarian cycle	Follicular phase		Luteal phase

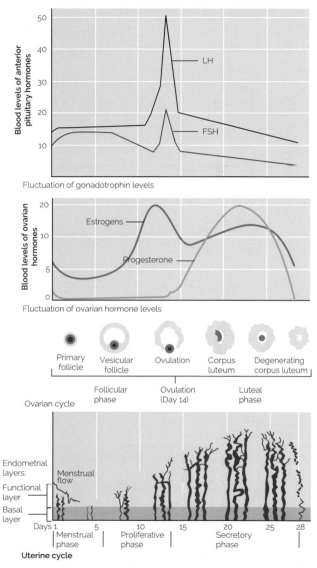

Figure 13.5: Menstrual cycle

Aging and the Reproductive System

Table 13.5: Aging and the reproductive system

Before puberty	The reproductive system appears to be dormant until a child reaches approximately 10 years of age. From this age, hormone-directed changes occur and the child goes through **puberty**
Puberty	The time in which a person develops secondary sexual characteristics and becomes able to reproduce

Male
Before puberty, which occurs around the age of 14, a boy has low levels of LH, FSH, and testosterone, and it is only at puberty that the levels of these hormones begin to increase under the influence of GnRH from the hypothalamus. Sustentacular cells in the testes mature and secrete testosterone, and spermatogenesis begins. Increased levels of testosterone bring about the development of secondary sexual characteristics, the enlargement of the reproductive glands, and both muscular and bone growth

Female
Before puberty, a girl has low levels of LH, FSH, and estrogens. However, under the influence of GnRH at the onset of puberty, LH and FSH stimulate the ovaries to produce estrogens and girls then develop secondary sexual characteristics and begin menstruating. This event is marked by **menarche**, which is a girl's first menses around the age of 12 years

Pregnancy in women	Once a woman has begun to menstruate, she is capable of falling pregnant. Pregnancy is the sequence of fertilization, implantation, embryonic growth, and fetal growth. An average pregnancy lasts 40 weeks from the first day of the woman's last menstrual period (approximately 38 weeks from conception) and is divided into three trimesters, each trimester being made up of approximately three months

Trimester 1, months 0–3
During the first trimester, the embryo implants and secures itself in the uterus. It grows from a single cell into a fully formed fetus in only 12 weeks, and by the end of the first trimester it has all its organs, muscles, limbs, and bones

Trimester 2, months 4–6
The fetus is now fully formed and just growing and maturing. During the second trimester it develops its individual fingerprints, its toe- and fingernails, its eyebrows and lashes, and a firm hand grip. It is in this trimester that it even starts grimacing and frowning

Table 13.5: (continued)

Trimester 3, month 7–birth	
During the third trimester, the fetus develops its sense of hearing, practices its breathing motions, and learns to focus and blink its eyes. It is fully formed and puts on a great deal of weight in the last few weeks	
Aging in women (menopause)	Although men are capable of reproducing well into old age, women are only capable of reproducing into their forties or fifties. At this age they go through **menopause**, which is the cessation of the menses. Menopause is often called the "change of life" and can be accompanied by a number of signs and symptoms, including hot flushes, headaches, thinning of hair and skin, sweating, vaginal dryness, decreased bone density, insomnia, weight gain, and mood swings. After menopause, a woman's reproductive organs begin to atrophy
Aging in men	From around the age of 55 years, testosterone levels begin to decline and men lose their muscular strength. Their sperm also become less viable and their libido decreases. However, healthy men are still able to reproduce into their eighties and sometimes even their nineties

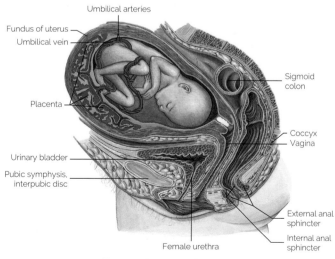

Figure 13.6: Uterus with a fetus

Theory in practice

The reproductive system is one of the most fascinating of all. It is the only system in the body that does not work continually from birth. Instead, it waits until puberty, when it bursts into action. It is the only system that is structurally and functionally different in men and women, and it is also the only system in which a unique type of cellular division occurs, creating cells with half the number of chromosomes of all other cells in the body.

It is also fascinating when you try to comprehend the numbers involved in this system. In males, over 300 million sperm cells mature every day, and 50–150 million sperm are present per milliliter of ejaculation. Yet only one sperm will get to fertilize an egg! In females, a baby girl is born with between 200,000 and 2 million oocytes in each ovary. However, many of these degenerate, and only about 400 will actually mature and be viable for ovulation.

Glossary

A

Abdomen	Region of the body between the diaphragm and pelvis
Abduction	Movement away from midline of body
Absorption	Uptake of digested nutrients into the bloodstream and lymphatic system
Acetylcholine	Neurotransmitter found in both the peripheral and central nervous systems
Acid mantle	Film of sebum and sweat on the surface of the skin that protects against bacteria
Actin	Protein that functions in muscle contraction
Action potential	Electrical charge that occurs on the membrane of a muscle cell in response to a nerve impulse
Active transport	Movement of a substance across a cellular membrane that involves the release of energy; takes place against a concentration gradient
Adduction	Movement toward the midline of the body
Adenosine triphosphate (ATP)	Main energy-transferring molecule in the body
Adipocyte	Fat cell
Adrenaline	*See* epinephrine
Adrenocorticotropic hormone (ACTH)	Hormone secreted by the anterior pituitary gland that stimulates and controls the adrenal cortex
Aerobic	Requiring oxygen
Afferent	Carrying toward a center
Afferent neurons	*See* sensory neurons
Agonist	*See* prime mover
Agranulocytes	Group of white blood cells that do not contain granules in their cytoplasm (includes lymphocytes and monocytes)
Aldosterone	Hormone secreted by the adrenal cortex that regulates the reabsorption of sodium and water in the kidneys
Alimentary canal	*See* gastrointestinal tract
Alveoli	Air sacs inside the lungs
Amphiarthroses	Slightly movable joints that permit a minimal amount of flexibility and movement
Anaerobic	Not requiring oxygen
Anagen	Active growing stage of the hair cycle
Anal canal	Last $\frac{3}{4}$–$1\frac{1}{5}$ in (2–3 cm) of the rectum that opens to the exterior

Anatomical position	Position in which the body is standing erect with the feet parallel, the arms hanging down by the side, and the face and palms facing forward
Anatomy	Study of the structure of the body
Antagonist	Muscle that opposes the movement caused by the prime mover, or agonist
Anterior	At the front of the body, in front of
Antibody	Specialized protein that is synthesized to destroy a specific antigen
Antidiuretic hormone (ADH)	Hormone released by the posterior pituitary gland that has an antidiuretic effect and raises blood pressure
Antigen	Any substance that the body recognizes as foreign
Antioxidant	Substance that combats or neutralizes free radicals
Apocrine gland	Type of sweat gland located in the armpits, pubic region, and the areolae of the breasts
Aponeurosis	Flat, sheet-like tendon that attaches muscles to bone, skin, or another muscle
Appendicular skeleton	Part of the skeleton consisting of the upper and lower limbs and their girdles
Appendix (vermiform)	Sac attached to the cecum of the large intestine
Aqueous humor	Fluid that nourishes the lens and cornea and helps produce intraocular pressure
Arachnoid	Middle meninx (covering) of the brain and spinal cord
Areola	Circular area of pigmented skin surrounding the nipple
Arrector pili muscles	Smooth muscles attached to hairs that contract to pull the hairs into a vertical position
Arteries	Vessels that usually carry blood away from the heart toward the tissues
Arterioles	Tiny arteries that deliver blood to capillaries
Arthrology	The study of joints
Articulation	Point of contact between two bones, commonly called a joint
Atony	Lack of muscle tone
Atrioventricular valves	Valves lying between the atria and ventricles that prevent the backflow of blood
Atrium	Receiving chamber of the heart
Atrophy	Wasting away of muscles
Auditory ossicles	Three tiny bones extending across the middle ear: the malleus, incus, and stapes
Auditory tube	*See* Eustachian tube
Auricle	The part of the ear that we see
Autonomic nervous system (ANS)	Part of the nervous system that controls all processes that are automatic or involuntary

Autorhythmic cells	Muscle or nerve cells that generate an impulse without an external stimulus; i.e., they are self-excitable
Avascular	Lacking blood vessels
Axial skeleton	Part of the skeleton comprising the bones found in the center of the body
Axilla	Armpit
Axon	Transmitting portion of a neuron
Axon terminal	Area found at the end of the axon that contains membrane-enclosed sacs called synaptic vesicles

B

B cells	Cells of the immune system that develop into plasma cells and are able to synthesise and secrete antibodies
Ball-and-socket joint	Synovial joint in which a ball-shaped bone fits into the cup-shaped socket of another bone
Baroreceptors	Sensory nerve endings that monitor blood pressure changes in the arteries and veins
Basophils	Type of white blood cell that contains histamine and is capable of ingesting foreign particles
Bile	Liquid produced by the liver that emulsifies fats; contains water, bile acids, bile salts, cholesterol, phospholipids, bile pigments, and some ions
Blood pressure	Force exerted by blood on the walls of a blood vessel
Bowman's capsule	*See* glomerular capsule
Brachial	Pertaining to the arm
Bradycardia	Slow heart rate
Brain stem	Continuation of the spinal cord that connects the spinal cord to the diencephalon; consists of the medulla oblongata, pons, and midbrain
Bronchus	Branchlike passageway inside the lungs
Brush border	Area in the small intestine composed of microvilli that contain digestive enzymes
Buccal cavity	Mouth
Bursa	Sac-like structure made of connective tissue, lined with a synovial membrane, and filled with synovial fluid

C

Calcitonin (CT)	Hormone secreted by the thyroid gland that lowers blood calcium levels
Calcitriol	Active form of vitamin D
Canaliculus	Very small channel or canal
Cancellous bone tissue	*See* spongy bone tissue
Capillary	Very small blood vessel that connects arterioles to venules

Carbohydrate	Organic compound composed of units of glucose that contain carbon, hydrogen, and oxygen
Cardiac cycle	All the events associated with a heartbeat
Cardiac muscle	Muscle forming most of the wall of the heart; composed of striated muscle fibers; involuntary
Cardiac output	Amount of blood pumped out of the heart by the left ventricle
Carpal	Pertaining to the wrist
Cartilage	Resilient, strong connective tissue that is less hard but more flexible than bone
Cartilaginous joint	Joint in which bone ends are held together by cartilage and do not have a synovial cavity between them
Catagen	The transitionary stage of the hair cycle
Catalyst	Substance that affects the rate of a chemical reaction without itself being changed by the reaction
Caudal	Away from the head, below
Cecum	$2\frac{2}{5}$ in (6 cm) long pouch of the large intestine that receives food from the small intestine via the ileocecal valve
Cellular respiration	Metabolic reaction that uses oxygen and glucose and produces energy in the form of ATP; also called oxidation
Central nervous system (CNS)	The brain and spinal cord
Centriole	Structure found near the nucleus of a cell that plays a role cell division
Centrosome	Area near the nucleus of a cell that contains centrioles and forms the mitotic spindle in dividing cells
Cephalad	Toward the head, above
Cephalic	Pertaining to the head
Cerebellum	Region of the brain located behind the medulla oblongata and pons; functions in producing smooth, coordinated movements as well as posture and balance
Cerebral cortex	Outer, most superficial layer of the cerebrum, consisting of gray matter
Cerebrospinal fluid (CSF)	Fluid that circles the CNS, protecting it and helping to maintain homeostasis
Cerebrum	Largest part of the brain and the area that gives us the ability to read, write, speak, remember, create, and imagine
Cerumen	Earwax
Cervical	Pertaining to the neck
Chemoreceptor	Receptor sensitive to chemicals
Choroid	Lining of most of the internal surface of the sclera
Chromatin	Mass of chromosomes all tangled together in a non-dividing cell
Chromosome	Threadlike structure found in the nucleus of a cell; carries the genes

Chyle	Fluid found inside the lacteals of the small intestine
Chyme	Semifluid contents of the stomach, consisting of partially digested food and gastric secretions
Cilia	Tiny hairlike projections on the surfaces of cells that move the cell or substances along the surface of the cell
Ciliary body	Part of the eye between the iris and choroid
Circumduction	Circular movement of the distal end of a body part; e.g., circling the shoulder joint
Cistern	Channel or tubule in a cell
Clitoris	Small cylindrical mass of erectile tissue and nerves that forms part of the female genitalia
Club hair	Fully grown hair that has detached from the hair bulb during the catagen stage of hair growth
Cochlea	Bony, spiral canal in the inner ear that resembles a snail's shell and houses the organ of Corti (for hearing)
Colon	Long tube that forms most of the large intestine
Compact bone tissue	Very hard, compact tissue with few spaces within it
Complement system	Group of proteins in the blood that helps antibodies during an immune response
Concentric contraction	Contraction toward a center that results in a movement that shortens the angle at a joint
Conductivity	Ability of cells, such as nerve or muscle cells, to move action potentials along their plasma membranes
Condyloid joint	Joint in which an oval protuberance at the end of a bone fits into an elliptical cavity of another bone
Cones	Photoreceptors that respond to color
Connective tissue	One of the basic tissue types in the body; consists of few cells in a large matrix and functions in support, storage, and protection
Contractility	Ability of muscles to contract and shorten
Cooper's ligaments	*See* suspensory ligaments
Cornea	Avascular, transparent coat that covers the iris; it is curved and helps focus light
Coronal plane	Plane that divides vertically into anterior and posterior portions
Coronary circulation	Circulation that supplies the muscles of the heart with blood
Corpus luteum	Gland formed after a Graafian follicle has discharged its ovum; secretes progesterone, estrogen, relaxin, and inhibin
Cortex	Outer layer of an organ
Corticotropin	*See* adrenocorticotropic hormone
Costal	Pertaining to a rib
Cranial	Toward the head, above, pertaining to the head
Cranium	Hard bones of the skull
Creatinine	Product of muscle activity

Cross-section	Division that divides horizontally into inferior and superior portions
Cutaneous	Pertaining to the skin
Cutaneous membrane	The skin; composed of an epidermis and dermis
Cuticle	Outer layer of cells of a hair or the epidermis of the skin or the base of the nail plate
Cytokinesis	Process by which a cell splits into two new cells during cellular division
Cytology	The study of cells
Cytoplasm	Cellular material inside the plasma membrane, excluding the nucleus
Cytosol	Thick, transparent, gel-like fluid inside a cell

D

Deep	Away from the surface of the body
Defecation	Process by which indigestible substances and some bacteria are eliminated from the body
Deglutition	Process of swallowing food
Demineralization	Process through which minerals such as calcium and phosphorus are lost from the bones
Dendrite	Receiving or input portion of a neuron
Dense connective (fibrous) tissue	Contains thick, densely packed fibers and fewer cells than loose connective tissue
Dentes	Teeth
Depolarization	Process by which an action potential is produced
Depression (of the shoulders or jaw)	Dropping the shoulders or jaw downward
Dermatology	Study of the skin
Dermis	Deep layer of the skin; composed of dense irregular connective tissue
Desquamation	Process by which the skin is shed
Diaphysis	Main, central shaft of a long bone
Diarthrosis	Freely movable joint that permits a number of different movements
Diastole	Relaxation of the heart muscle during the cardiac cycle
Diencephalon	Region of the brain that lies above the brain stem, enclosed by the cerebral hemispheres; contains the thalamus, hypothalamus, and epithalamus and has a number of different functions, including housing the pituitary and pineal endocrine glands
Diffusion	Movement of substances from areas of high concentration to areas of low concentration
Digestion	Process by which large molecules of food are broken down into smaller molecules that can enter cells

Digit	Finger or toe
Diploid	Having two complete sets of chromosomes per cell (i.e., 46)
Distal	Farther from its origin or point of attachment of a limb
Distal convoluted tubule	Area furthest from the glomerular capsule; section of the tube where fine-tuning of the filtrate occurs
Diuretic	Substance that increases urine production
Dopamine	Type of neurotransmitter
Dorsal	At the back of the body, behind
Dorsiflexion	Pulling of the foot upward toward the shin, in the direction of the dorsum
Duodenal glands (Brunner's glands)	Glands in the duodenum that secrete an alkaline mucus
Duodenum	First segment of the small intestine; connects the stomach to the ileum
Dura mater	Outer covering of the brain

E

Eardrum (tympanic membrane)	Very thin, semi-transparent membrane between the auditory canal and the middle ear; when sound waves hit it, it vibrates, passing the sound waves on to the middle ear
Eccentric contraction	Contraction away from the center
Eccrine gland	Sweat gland; distributed around the body
Efferent	Carrying away from a center
Efferent neuron	*See* motor neuron
Elasticity	Ability of a tissue to return to its original shape after stretching, contracting, or extending
Electrolyte	Charged particle (ion) that conducts an electrical current in an aqueous solution
Elevation (of the shoulders or jaw)	Lifting the shoulders or jaw upward
Ellipsoid joint	*See* condyloid joint
Enamel	Extremely hard substance that protects the teeth from being worn down and acts as a barrier against acids
Endocardium	Thin, smooth lining of the inside of the heart
Endocrine glands	Ductless glands that secrete substances into the extracellular space around their cells; these secretions then diffuse into blood capillaries and are transported by the blood to target cells located throughout the body
Endocrinology	Study of the endocrine glands and the hormones they secrete
Endometrium	Mucous membrane lining of the uterus
Endomysium	Connective tissue that surrounds each individual muscle fiber

Endoplasmic reticulum	Network of fluid-filled cisterns within a cell that provides a large surface area for chemical reactions and also transports molecules within the cell
Endosteum	Membrane that lines the medullary cavity of bones
Enteroendocrine cell	Specialized cell that secretes hormones into the intestinal glands
Enzyme	Protein that speeds up a chemical reaction without itself being used up in the reaction
Eosinophils	Type of white blood cell that can destroy certain parasitic worms, phagocytize antigen–antibody complexes, and combat the effects of some inflammatory chemicals
Epicardium	Outer layer of the heart wall; also called the visceral layer of the serous pericardium
Epidermis	Superficial, outer layer of the skin
Epididymis	The organ lying along the posterior border of each testis; composed of a series of coiled ducts; the site of sperm maturation
Epimysium	The outermost layer of connective tissue that encircles an entire muscle
Epinephrine	Hormone secreted by the adrenal medulla that functions in the fight-or-flight response
Epiphyseal plate	A layer of hyaline cartilage in a growing bone that allows the diaphysis to grow in length
Epiphysis	The end of a long bone
Epithelium	Basic tissue type; forms glands, lines internal cavities and vessels, and is the superficial layer of the skin
Equilibrium	Balance
Erythrocyte	Red blood cell, contains a protein called hemoglobin, which transports oxygen in the blood
Erythropoietin	Hormone secreted by the kidneys that stimulates the production of red blood cells
Essential fatty acid	Fats that are vital for the proper functioning of the body
Estrogens	Hormones secreted by the ovaries that stimulate the development of feminine secondary sex characteristics and, together with progesterone, regulate the female reproductive cycle
Eustachian tube	Tube that connects the middle ear with the upper portion of the throat; equalizes the middle-ear-cavity pressure with the external atmospheric pressure
Eversion	Turning the sole of the foot outward
Excitability	Ability of muscle or nerve cells to respond to stimuli
Excretion	Elimination of waste products
Exocrine glands	Glands that secrete substances into ducts that carry these substances into body cavities or to the outer surface of the body

Extensibility	Ability of muscles to extend and lengthen or stretch
Extension	Straightening movement in which a body part is restored to its anatomical position after being flexed
External auditory canal	Curved tube that carries sound waves from the auricle to the eardrum
External nares	Openings to the nose; commonly called the nostrils
External respiration (pulmonary respiration)	Gaseous exchange between lungs and blood; in external respiration, the blood gains oxygen and loses carbon dioxide

F

Facilitated diffusion	Diffusion in which substances are helped across the plasma membrane by channel or transporter proteins within the membrane
Falciform ligament	Fold of the peritoneum that binds the liver to the anterior abdominal wall and diaphragm and separates the two principal lobes of the liver
Fallopian tubes	Two thin tubes running from the ovaries to the uterus
Fascia	Connective tissue that surrounds and protects organs, lines walls of the body, holds muscles together, and separates muscles
Fascicle	Bundle of 10–100 muscle fibers
Fatigue (of muscles)	A muscle's inability to respond to stimulus or maintain contractions
Feces	Waste material of the digestive system that is eliminated through the anus
Fertilization	Union and fusion of an ovum and a spermatozoon to form a zygote
Fetus	Unborn child in the uterus from the eighth week of development until birth
Fibrous joints	Joints in which bone ends are held together by fibrous connective tissue with no synovial cavity between them
Filtration	Movement of a liquid through a membrane or filter
Fixator	Muscle that helps the prime mover by stabilizing and preventing unnecessary movements in surrounding joints
Flagella	Long, whiplike extensions of the cell membrane of certain cells such as sperm or bacteria; they move the cell
Flat bones	Thin bones consisting of a layer of spongy bone enclosed by layers of compact bone
Flexion	Bending of a joint in which the angle between articulating bones decreases; the opposite of extension
Follicle-stimulating hormone (FSH)	Hormone secreted by the anterior pituitary gland that stimulates the development of ova and sperm
Formed elements	Cells and cell fragments found in blood

Free radical   Highly unstable, reactive molecule that damages cells

Frontal plane   *See* coronal plane

G

Gamete   Sex cell (ovum or spermatozoon)

Ganglion   Bundle or knot of nerve cell bodies

Gastric juice   Substance secreted by the gastric glands in the stomach, containing water, hydrochloric acid, intrinsic factor, pepsinogen, and gastric lipase

Gastric lipase   Enzyme that acts on lipids in the stomach, breaking down triglycerides into fatty acids and monoglycerides

Gastrin   Hormone produced by the stomach that stimulates gastric secretions

Gastrointestinal (GI) tract   Tube that runs from the mouth to the anus in which digestion and absorption take place

Gene   Basic unit of genetic material

Germinal matrix   Region of the nail where cell division takes place and growth occurs

Gingivae   Gums

Gliding joint   Joint in which two flat surfaces meet

Glomerular (Bowman's) capsule   Cuplike structure surrounding the glomerulus in a kidney's nephron; forms the closed end of the renal tubule

Glomerulus   Knotted network of capillaries in a kidney's nephron

Glucagon   Hormone produced by the pancreas that raises blood glucose levels

Glucocorticoids   Group of hormones secreted by the adrenal cortex that stimulate metabolism, help the body resist long-term stressors, control the effects of inflammation, and depress immune responses

Glucose   Sugar that is the major energy source for all cells

Gluteal   Pertaining to the buttocks

Glycogen   Carbohydrate consisting of subunits of glucose; the main form in which carbohydrates are stored in the body

Glycolysis   Cellular process through which glucose is split into pyruvic acid and ATP

Goblet cells   Cells that secrete mucus

Golgi apparatus   *See* Golgi complex

Golgi complex   Cellular structure located near the nucleus that processes, sorts, and packages proteins and lipids for delivery to the plasma membrane; also forms lysosomes and secretory vesicles

Gonad   Male or female reproductive organ in which sex cells are produced

Graafian follicle   Large, fluid-filled follicle that ruptures and releases an ovum during ovulation

Granulocytes	Group of white blood cells containing granules in their cytoplasm (includes neutrophils, eosinophils, and basophils)
Gustation	Sense of taste

H

Hair matrix	Ring of cells that divide to create hair
Hair root	Portion of the hair that penetrates into the dermis
Hair shaft	Superficial end of the hair that projects from the surface of the skin; commonly called the hair strand
Haploid cell	Cell with a single set of chromosomes (23 chromosomes)
Haustra	Pouches on the external surface of the colon
Haversian system	*See* osteon
Heart rate	Number of times the heart beats in one minute
Hemoglobin	Protein found in red blood cells that transports oxygen and gives the blood cells their red color
Hemopoiesis	Production of blood cells and platelets
Hepatocyte	Specialized cell found in the liver; has many metabolic functions
Hinge joint	Joint in which the convex surface of a bone fits into the concave surface of another bone
Histology	Study of tissues
Homeostasis	Process by which the body maintains a stable internal environment
Hormone	Chemical messenger regulating cellular activity, produced by an endocrine gland and transported in the blood
Human growth hormone (hGH)	Hormone secreted by the anterior pituitary gland that stimulates growth and regulates metabolism
Hydrophilic	Water-loving
Hydrophobic	Water-hating
Hydroxyapatite	Form of calcium phosphate found in bones and teeth
Hymen	Thin membrane that partially covers the opening of the vagina
Hyperextension	Occurs when a body part extends beyond its anatomical position
Hypersecretion	Over- or excessive secretion
Hyposecretion	Under-secretion

I

Ileum	Longest segment of the small intestine, which receives food from the jejunum and passes it into the large intestine
Immunity	Ability of the body to resist infection
Inclusions	Diverse group of substances that are temporarily produced by some cells
Inferior	Away from the head, below

Inflammation	Body's response to tissue damage
Ingestion	Process of taking food into the mouth
Inguinal	Pertaining to the groin
Inner ear	System of cavities and ducts that contains the organs of hearing and balance; also called the labyrinth
Insertion	Point where a muscle attaches to the moving bone of a joint
Insulin	Hormone produced by the pancreas that lowers blood glucose levels
Integumentary system	System of the skin and its derivatives (hair, nails, and cutaneous glands)
Internal nares	Openings that connect the nasal cavity to the pharynx
Internal respiration	Gaseous exchange between the blood and tissue cells; in internal respiration the blood loses oxygen and gains carbon dioxide
Interstitial endocrinocytes	Cells in the testes that secrete testosterone; also called Leydig cells
Intestinal juice	Clear yellow fluid that has a slightly alkaline pH and contains water and mucus; helps bring nutrient particles into contact with the microvilli
Intrinsic factor	Substance produced by the stomach and necessary for the absorption of vitamin B12 from the ileum
Inversion	Turning the sole of the foot inward
Ion	Electrically charged molecule or atom
Iris	Colored portion of the eye, suspended between the cornea and lens; contains the pupil
Irregular bones	Bones that have complex shapes and varying amounts of compact and spongy tissues
Irritability	Ability to respond to a stimulus and convert it into an impulse
Islets of langerhans	Clusters of endocrine cells located in the pancreas that secrete hormones including insulin and glucagon
Isometric contraction	Contraction in which the muscle contracts but does not shorten and no movement is generated
Isotonic contraction	Contraction in which muscles shorten and create movement while the tension in the muscle remains constant

J

Jaundice	Yellowing of skin or whites of eyes due to high levels of bilirubin
Jejunum	Portion of the small intestine between the duodenum and the ileum

K

Keratinization	Process in which cells die and become full of the protein keratin
Keratinocytes	Cells that produce keratin; 95% of the cells of the epidermis are keratinocytes

Kinesiology	Study of the motion of the body
Kinetic energy	Energy of motion

L

Labia	Lips
Labia majora	Two longitudinal folds of skin that extend inferiorly and posteriorly from the mons pubis of the female genitalia
Labia minora	Two smaller folds of skin running medially to the labia majora of the female genitalia
Labyrinth	*See* inner ear
Lacrimal gland	Gland that secretes tears
Lactase	Brush-border enzyme that breaks down lactose into glucose and galactose
Lactation	Secretion of milk by mammary glands
Lacteals	Specialized lymphatic vessels found in the villi of the small intestine
Lactiferous	Pertaining to the breasts
Lactogenic hormone	*See* prolactin
Lamellae	Concentric rings of calcified matrix found in compact bone
Lamellated corpuscles	Nerve endings that are sensitive to pressure; also called Pacinian corpuscles
Langerhans cell	A cell that functions in skin immunity
Lanugo hair	Soft hair that begins to cover a fetus from the third month of pregnancy; usually shed by the eighth month of pregnancy
Larynx	Short passageway between the laryngopharynx and the trachea; commonly called the voice box
Lateral	Away from the midline, on the outer side
Lens	Transparent structure of the eye, located behind the iris and responsible for fine-tuning of focusing
Leucocyte	White blood cell; functions primarily in protecting the body against foreign microbes and in immune responses
Ligament	Tough band of connective tissue that attaches bones to bones
Limbic system	Region of the brain that controls the emotional and involuntary aspects of behavior and also functions in memory
Lingual frenulum	Fold of mucous membrane that secures the tongue to the floor of the mouth
Lingual lipase	Enzyme in the mouth that begins the breakdown of lipids from triglycerides into fatty acids and glycerol
Lipid	A fat; fats are organic compounds composed of carbon, hydrogen, and oxygen and they are usually insoluble in water
Long bones	Bones that have a greater length than width and usually contain a longer shaft with two ends
Longitudinal plane	Plane that divides vertically into right and left sides

Loop of Henle	Portion of the renal tubule of a kidney nephron
Loose connective tissue	Tissue that consists of two or more layers of cells; it is durable and functions in protecting underlying tissues in areas of wear and tear
Lumbar	Lower back region between the thorax and pelvis
Lumen	Hollow space within a tube-like structure, such as an artery, vein, or intestine
Lunula	Crescent-shaped white area at the proximal end of the nail plate
Luteinizing hormone (LH)	Hormone secreted by the anterior pituitary gland that stimulates ovulation, formation of the corpus luteum, and secretion of estrogens and progesterone in females; also stimulates production of testosterone in males
Lymph	Clear, straw-colored fluid derived from interstitial fluid
Lymphocyte	Type of white blood cell involved in immunity; B cells and T cells are types of lymphocyte
Lymphoid tissue	Tissue where lymphocytes and antibodies are produced; found in lymph nodes, the tonsils, the thymus, the spleen, and as diffuse cells
Lysosomes	Cellular vesicles containing powerful digestive enzymes that can break down and recycle many different molecules
Lysozyme	Enzyme found in certain body secretions such as tears and saliva; catalyzes the breakdown of the cell walls of certain bacteria

M

Macrophage	Scavenger cell that engulfs and destroys microbes
Maltase	Brush-border enzyme that breaks down maltose into glucose
Mammary glands	Two modified sudoriferous glands; commonly called the breasts
Marrow cavity	*See* medullary
Mast cells	Large cells in connective tissue that release substances such as histamine during inflammation
Mastication	Process of chewing food
Meatus	*See* external auditory canal
Medial	Toward the midline, on the inner side
Median line	Imaginary line through the middle of the body
Mediastinum	Space in the thorax containing the aorta, heart, trachea, esophagus, and thymus gland; found between the two pleural sacs
Medulla	Inner layer of an organ
Medullary	Space within the diaphysis of a bone; contains yellow bone marrow and is also called the marrow cavity
Meiosis	Reproductive cell division in which four haploid daughter cells are produced

Meissner's corpuscles	Nerve endings that are sensitive to touch
Melanocytes	Melanin-producing cells
Melanocyte-stimulating hormone (MSH)	Hormone secreted by the anterior pituitary gland; exact actions are unknown, but can cause darkening of the skin
Melatonin	Hormone secreted by the pineal gland that causes sleepiness
Membrane	Thin, flexible sheet made up of different tissue layers; membranes cover surfaces, line body cavities, and form protective sheets around organs
Meninges (sing. meninx)	Three connective tissue membranes that enclose the brain and spinal cord
Meniscus	Pad of fibrocartilage that lies between the articular surfaces of bones
Menopause	When a woman stops menstruating and ovulating and can no longer bear children
Menses	*See* menstruation
Menstrual cycle	Cycle in which the endometrium of the uterus is prepared for the arrival of a fertilized ovum
Menstruation	Cyclical, periodic discharge of menstrual flow from the uterus; contains blood, tissue fluid, mucus, and epithelial cells derived from the endometrium
Merkel cells	Cells only found in the stratum basale of hairless skin and attached to keratinocytes; they make contact with nerve cells to form Merkel discs, which function in the sensation of touch
Metabolism	Changes that take place within the body to enable its growth and function
Microvilli	Tiny, membrane-covered projections extending into the small intestine and increasing surface area for absorption
Micturition	Urination
Middle ear	Small, air-filled cavity found between the outer ear and the inner ear; contains the three auditory ossicles
Midline	*See* median line
Mineralocorticoids	Group of hormones secreted by the adrenal cortex that regulate the mineral content of the blood
Mitochondria	Powerhouses of the cell where ATP is generated through the process of cellular respiration
Mitosis	Cellular reproduction in which a mother cell divides into two daughter cells, each containing the same genes as the mother cell
Mitral valve	Left atrioventricular valve
Mixed nerve	Nerve containing both sensory and motor fibers
Monocyte	Type of white blood cell
Mons pubis	Elevation of adipose tissue that cushions the pubic symphysis

Motor neuron   Neuron that conducts impulses from the CNS to muscles and glands

Mucosa   *See* mucous membrane

Mucosa-associated lymphoid tissue (MALT)   Concentrations of lymphatic tissue that are strategically positioned to help protect the body from pathogens that have been inhaled, digested, or have entered via external openings

Mucous membrane   Membrane that lines body cavities that open directly to the exterior; they are wet membranes whose cells secrete mucus

Muscle tissue   Tissue composed of elongated cells that are able to shorten (contract) to produce movement

Muscularis   Muscular layer or coat of an organ

Myelin sheath   Sheath that protects and insulates a neuron; composed of a white fatty substance called myelin

Myocardium   Middle layer of the heart wall; composed of cardiac muscle tissue and contracts to pump blood

Myofiber   Muscle fiber

Myofibrils   Long, threadlike organelles, the contractile elements of a skeletal muscle fiber

Myofilaments   Filaments found inside myofibrils; there are two types: actin/thin and myosin/thick filaments

Myogenic rhythm   Inherent rhythmicity of certain muscles that does not rely on nervous stimulation; e.g., in cardiac muscle

Myoglobin   Protein that binds with oxygen and carries it to muscle cells

Myology   Study of muscles

Myosin   Protein that functions in muscle contraction

N

Nail bed   Area that lies directly beneath the nail plate and secures the nail to the finger or toe

Nail free edge   The part of the nail that extends past the end of the finger or toe; also called the distal edge

Nail grooves   Grooves on the sides of the nail that guide it up the fingers and toes

Nail mantle   The skin that lies directly above the germinal matrix of the nail

Nail plate   The visible body of the nail

Nail wall   The skin that covers the sides of the nail plate and protects the nail grooves

Nasal conchae   Bony shelves projecting from the lateral walls of the nasal cavity

Natural killer cells (NK cells)   Type of lymphocyte that can kill a variety of microbes as well as some tumor cells

Nephron   Functional unit of the kidney where filtration occurs

Nerve fiber   Term referring to the processes that project from a nerve body; e.g., dendrites and axons

Nerve tissue	Tissue made up of neurons and neuroglia; found in the brain, spinal cord, and nerves and functions in communication
Neurofibril node	*See* node of Ranvier
Neuroglia	Supporting cells that insulate, support, and protect neurons
Neurolemmocytes	*See* Schwann cells
Neurology	Study of the nervous system
Neuron	Nerve cell responsible for the sensory, integrative, and motor functions of the nervous system
Neurotransmitter	Chemical that transmits impulses across synapses from one nerve to another
Neutrophil	Type of white blood cell that engulfs and digests foreign particles and removes waste through phagocytosis
Node of Ranvier	Gap along a myelinated nerve fiber
Noradrenaline	*See* norepinephrine
Norepinephrine	Hormone secreted by the adrenal medulla that functions in the fight-or-flight response
Nucleic acid	Organic compound composed of nucleotides: DNA or RNA
Nucleolus	Spherical body inside the nucleus made up of protein, some DNA, and RNA
Nucleus	Structure in a cell that controls all cellular structure and activities and contains most of the genes

O

Oblique plane	Plane that divides at an angle between a transverse plane and a frontal or sagittal plane
Occipital	Pertaining to the back of the head
Olfaction	Sense of smell
Omentum	Double layer of the peritoneum; covers and links the abdominal organs
Onyx	Nail
Oocyte	Immature ovum
Oogenesis	Production of mature ova in the ovaries
Ophthalmic	Pertaining to the eye
Organelle	Little organ of the cell
Organic compound	Compound containing carbon
Origin	Point where a muscle attaches to the stationary bone of a joint
Osmoreceptor	Receptor sensitive to a decrease in water or an increase in solutes in the blood
Osmosis	Diffusion of water through a selectively permeable membrane from an area of lower solute concentration to an area of higher solute concentration
Osseous tissue	Bone tissue; an exceptionally hard connective tissue that protects and supports other organs of the body

Ossification	Process of bone formation
Osteoblast	Cell that secretes collagen and other organic components to form bones
Osteoclast	Cell found on the surface of bones that destroys or resorbs bone tissue
Osteocyte	Mature bone cell that maintains the daily activities of bone tissue; derived from osteoblasts and the main cells found in bone tissue
Osteology	Study of the structure and function of bones
Osteon	Basic unit of structure of an adult compact bone; consists of a system of interconnecting canals called Haversian canals
Osteoprogenitor cell	Stem cell derived from mesenchyme (the connective tissue found in an embryo) that has the ability to become an osteoblast
Outer ear	External region of the ear that collects and channels sound waves inward; composed of the auricle, external auditory canal, and eardrum
Ova	Mature female sex cells (commonly called eggs)
Ovarian cycle	Cycle in which an oocyte matures until it is ready for ovulation
Ovulation	Event occurring around day 14 of the 28-day cycle; involves the rupture of the mature Graafian follicle and the release of an ovum into the pelvic cavity
Oxidation	*See* cellular respiration
Oxytocin (OT)	Hormone released by the posterior pituitary gland that stimulates contraction of the uterus during labor and stimulates the milk let-down reflex during lactation

P

Pancreatic amylase	Enzyme present in pancreatic juice that completes the breakdown of starches and glycogen
Pancreatic islets	*See* islets of Langerhans
Pancreatic juice	Pancreatic secretion composed of mostly water, some salts, sodium bicarbonate, and some enzymes
Pancreatic lipase	Enzyme present in pancreatic juice; breaks down triglycerides into fatty acids and monoglycerides
Paneth cells	Specialized cells that secrete a bactericidal enzyme called lysozyme into the small intestine
Papillae	Small projections covering the tongue; some papillae house the taste buds
Papillary layer	Undulating membrane that makes up approximately $1/5$ of the thickness of the dermis; composed of areolar connective tissue and fine elastic fibers and has nipple-shaped fingerlike projections called papillae
Parasympathetic nervous system	Nervous system that opposes the actions of the sympathetic nervous system by inhibiting activity, thus conserving energy

Parathormone (PTH)	Hormone secreted by the parathyroid glands that increases blood calcium and magnesium levels, decreases blood phosphate levels, and promotes formation of calcitriol by the kidneys
Parathyroid hormone	*See* parathormone
Parietal	Relating to the wall of the body or any of its cavities
Pepsin	Enzyme in the stomach that begins the breakdown of proteins
Pepsinogen	Enzyme precursor in the stomach that is converted into pepsin in the acidic environment of gastric juice
Peptidases	Brush-border enzymes that complete the breakdown of proteins into amino acids
Pericardium	Membranous sac that surrounds and protects the heart
Perimysium	Connective tissue that surrounds bundles of 10–100 muscle fibers
Perineum	Diamond-shaped area that contains the external genitals and the anus; located between the thighs and buttocks and present in both males and females
Periosteum	Connective tissue membrane that covers bones
Peripheral	At the surface or outer part of the body
Peripheral nervous system (PNS)	Part of the nervous system connecting the rest of the body to the central nervous system (brain and spinal cord)
Peristalsis	Involuntary wave-like movement that pushes the contents of the gastrointestinal tract forward
Peritoneum	Large serous membrane lining the abdominal cavity
Peroxisomes	Cellular vesicles containing enzymes that detoxify any potentially harmful substances in the cell
Peyer's patches	Patches of mucosa-associated lymphoid tissue located in the lining of the small intestine
Phagocytosis	Engulfment and digestion of foreign particles by phagocytes
Phagocyte	Cell that can engulf and digest microbes; phagocytes include macrophages and some types of white blood cells
Phalanges	Bones of the fingers and toes
Pharynx	Funnel-shaped tube whose walls are made up of skeletal muscles lined by mucous membrane and cilia; commonly called the throat
Photoreceptors	Specialized cells that convert light into nerve impulses
Physiology	Study of the functions of the body
Pia mater	Thin inner covering of the brain; dips into all the folds and spaces of the brain tissue
Pivot joint	Joint in which a rounded/pointed surface of a bone fits into a ring-shaped bone

Plane	Imaginary flat surface that divides the body or organs into parts
Plane joint	*See* gliding joint
Plantar	Pertaining to the sole of the foot
Plantar flexion	Pointing of the foot downward, in the direction of the plantar surface
Plasma	Liquid portion of blood
Plasma cell	Cell that develops from a B cell (type of lymphocyte) and produces antibodies
Plasma membrane	Barrier that surrounds a cell and regulates the movement of all substances into and out of it
Platelet	*See* thrombocyte
Plexus	Network of nerves or blood vessels
Polyuria	Production of more than 3 liters of urine per day
Popliteal	Relating to the hollow space behind the knee
Posterior	At the back of the body, behind
Postovulatory phase	Phase starting 14 days after ovulation in which the endometrium awaits the arrival of a fertilized ovum
Pregnancy	Sequence of fertilization, implantation, embryonic growth, and fetal growth
Preovulatory phase	Time between menstruation and ovulation; in a 28-day cycle it can vary from 6 to 13 days in length and is the time when a mature Graafian follicle forms and the endometrium proliferates
Prime mover	Muscle responsible for causing a movement
Process	Bony projection or prominence
Progesterone	Hormone secreted by the ovaries that, together with estrogens, regulates the female reproductive cycle and helps maintain pregnancy
Prolactin (PRL)	Hormone secreted by the anterior pituitary gland that stimulates the secretion of milk from the breasts
Proliferative phase	*See* preovulatory phase
Pronation	Movement involving turning the palm posteriorly or inferiorly
Proprioceptor	Specialized nerve receptor located in muscles, joints, and tendons that provides sensory information regarding body position and movements
Prostate gland	Gland surrounding the prostatic urethra; secretes a milky, slightly acidic fluid that contributes to sperm mobility and viability
Protein	Organic compound made up of amino acids and containing carbon, hydrogen, oxygen, and nitrogen; the main building material of cells
Protraction	Drawing the shoulders or jaw forward
Proximal	Closer to its origin or point of attachment of a limb

Proximal convoluted tubule	Portion of the renal tubule closest to the glomerular capsule of a kidney nephron; where the reabsorption of most substances takes place
Puberty	Time at which a person develops secondary sexual characteristics and becomes able to reproduce
Pulmonary circulation	Circulatory system in which the right side of the heart receives deoxygenated blood from the body and pumps it to the lungs where it is oxygenated
Pulmonary respiration	*See* external respiration
Pulmonary ventilation	Process in which air is inspired or breathed into the lungs and expired or breathed out of the lungs
Pupil	Hole in the center of the iris through which light enters the eye

Q

Quadrant	Region of the abdominopelvic cavity

R

Rectum	Last portion of the gastrointestinal tract
Reflex	Automatic response to a stimulus
Remodeling	Process through which new bone tissue replaces old, worn-out, or injured bone tissue
Renal	Pertaining to the kidneys
Renin	Enzyme produced by the kidneys that functions in raising blood pressure
Rennin	Enzyme found only in the stomachs of infants; begins the digestion of milk by converting the protein caseinogen into casein
Respiration	Exchange of gases between the atmosphere, blood, and cells
Retina	Innermost layer of the wall of the eyeball; consists of a non-visual pigmented portion and a neural portion
Retraction	Drawing the shoulders or jaw backward
Ribosomes	Tiny granules that are sites of protein synthesis in the cell
Rima glottidis	Vocal folds
Rods	Photoreceptors that respond to different shades of gray only
Rotation	Movement of a bone in a single plane around its longitudinal axis
Rugae	Folds in the mucous lining of a hollow organ; found in the stomach and vagina

S

Saddle joint	Joint in which a surface shaped like the legs of a rider fits into the saddle-shaped surface of another bone
Sagittal plane	Plane that divides vertically into right and left sides

Salivary amylase	Enzyme present in saliva that begins the breakdown of large carbohydrate molecules
Sarcolemma	Plasma membrane of a muscle fiber
Sarcomere	Basic functional unit of a skeletal muscle
Sarcoplasm	Cytoplasm of a muscle fiber
Schwann cells	Cells that wrap around the axon of a neuron to form a myelin sheath
Sclera	White of the eye; made up of dense connective tissue; protects the eyeball and gives it its shape and rigidity
Scrotum	Sac of loose skin and superficial fascia in which lie the testes
Sebaceous oil gland	Exocrine gland that secretes sebum; usually associated with a hair follicle
Sebum	Oily substance secreted by sebaceous glands
Secretion	Substance released from a gland cell; secretions are usually useful substances as opposed to waste products
Secretory phase	*See* postovulatory phase
Segmentation	Main movement in the small intestine; involves localized contractions that move food back and forth
Semen (seminal fluid)	Fluid containing sperm and a mixture of fluids secreted by the reproductive glands
Semicircular canals	Three semicircular canals that project from the vestibule in the ear and contain receptors for equilibrium
Semilunar valves	Valves lying between the ventricles and the arteries that prevent the backflow of blood
Seminal fluid	*See* semen
Seminal vesicles	Paired, pouch-like structures located at the base of the bladder that secrete a viscous alkaline fluid
Seminiferous tubules	Tightly coiled tubules located in the testes, where sperm are formed
Sensory neurons (afferent neurons)	Neurons that conduct impulses from sensory receptors to the CNS
Serous membrane	Membrane lining body cavities that do not open directly to the exterior and covering the organs that lie within those cavities
Sesamoid bones	Oval bones that develop in tendons where there is considerable pressure
Short bones	Cube-shaped bones that are nearly equal in length and width
Simple epithelium	Single layer of cells; usually very thin and functions in absorption, secretion, and filtration
Sinoatrial node	The heart's pacemaker (SA node)
Skeletal muscle tissue	Muscle tissue attached to bones that is composed of long, cylindrical fibers that are striated and under voluntary control
Smooth muscle tissue	Muscle tissue that contains non-striated (smooth) fibers and is regulated by the autonomic nervous system

Solvent	Medium (usually a liquid) in which substances (solutes) can be dissolved
Somatic cell	Any cell except the reproductive cells
Somatic cell division	Process of nuclear division called mitosis; a single diploid parent cell duplicates to produce two identical daughter cells
Somatic nervous system (voluntary nervous system)	Part of the nervous system that allows us to control our skeletal muscles; also called the voluntary nervous system
Somatostatin	Hormone secreted by the pancreas that inhibits insulin and glucagon release
Somatotropin	*See* human growth hormone
Spermatogenesis	Production of sperm in the testes
Spermatozoa	Mature male sex cells (commonly called sperm)
Spinal nerves	Nerves emerging from the spinal cord that carry impulses to and from the rest of the body
Spongy bone tissue	Light bone tissue with many spaces within it and a sponge-like appearance; does not contain osteons
Stratified epithelium	Consists of two or more layers of cells; durable; protects underlying tissues in areas of wear and tear
Stratum basale	Deepest layer of the epidermis and the base from where new cells germinate or sprout
Stratum corneum	Outermost layer of the skin, consisting of dead, tough cells
Stratum functionalis	Deep layer of the endometrium of the uterus; shed during menstruation
Stratum germinativum	*See* stratum basale
Stratum granulosum	Epidermal layer of degenerating cells that are becoming increasingly filled with little grains or granules of keratin
Stratum lucidum	Epidermal waterproof layer of dead, clear cells
Stratum spinosum	Epidermal layer of prickly cells that are beginning to go through the process of keratinization
Striated	Having the appearance of light and dark bands or striations
Subcutaneous	Beneath the skin
Substrate	Substance on which an enzyme acts
Sucrase	Brush-border enzyme that breaks down sucrose into glucose and fructose
Sudoriferous gland	Gland that excretes sweat onto the surface of the skin
Superficial	Toward the surface of the body
Superior	Toward the head, above
Supination	Movement involving turning the palm anteriorly or superiorly
Suspensory (Cooper's) ligaments	Strands of connective tissue that support a breast

Sympathetic nervous system	Part of the nervous system that reacts to changes in the environment by stimulating activity, thereby using energy
Synapse	Gap at the end of a nerve fiber that an impulse crosses to pass from one neuron to the next
Synaptic vesicle	Sac that stores neurotransmitters and is located in a synaptic end bulb at the distal end of an axon terminal
Synarthrosis	Immovable joint
Synergist	Muscle that helps the prime mover
Synovial joint	Freely movable joint in which a cavity is present between the articulating bones
Synovial membrane	Membrane composed of areolar connective tissue that lines freely movable joints and secretes synovial fluid
Systemic circulation	Type of circulation in which the left side of the heart receives oxygenated blood from the lungs and pumps it to the rest of the body
Systole	Contraction of the heart muscle during the cardiac cycle

T

T cell	Lymphocyte that matures in the thymus gland and functions in cell-mediated immunity
Tachycardia	Abnormally fast heart rate
Taeniae coli	Three thickened bands of longitudinal smooth muscle running the length of the colon
Taste bud	A receptor for taste
Telogen	Resting phase of the hair cycle in which the follicle is inactive until stimulated to develop another hair
Tendon	Strong cord of dense connective tissue that attaches muscles to bones, to the skin, or to other muscles
Terminal hair	Hair found on the head, eyebrows, eyelashes, under the arms, and in the pubic area
Testes	Gonads of the male reproductive system
Testosterone	Hormone secreted by the testes that stimulates the development of masculine secondary sex characteristics and libido
Thermogenesis	Generation of heat in the body
Thorax	The chest
Thrombocyte	Type of white blood cell that functions in hemostasis and blood clotting
Thymosin	Hormone secreted by the thymus gland that promotes the growth of T cells
Thyroid hormone	Hormone secreted by the thyroid gland that functions in metabolism, growth, and development
Thyroid-stimulating hormone (TSH)	Hormone secreted by the anterior pituitary gland that controls the thyroid gland

Thyrotropin	*See* thyroid-stimulating hormone
Thyroxine	*See* thyroid hormone
Tissue respiration	*See* internal respiration
Tone (tonus)	Partial contraction of a resting muscle
Trachea	Long, tubular passageway that transports air from the larynx into the bronchi; commonly called the windpipe
Tract	Bundle of nerve fibers that is not surrounded by connective tissue
Transverse plane	Plane that divides horizontally into inferior and superior portions
Tricuspid valve	Right atrioventricular valve
Trimester	Period of three months; pregnancy is divided into three trimesters
Trypsin	Enzyme secreted by the pancreas that continues the breakdown of proteins
Tympanic membrane	*See* eardrum

U

Unguis	Nail
Urea	Main product of protein metabolism
Uric acid	Product of nucleic acid metabolizm
Urobilinogen	Bile pigment derived from the breakdown of hemoglobin
Urology	Study of the urinary system
Uterine cycle	*See* menstrual cycle
Uterus	The womb
Uvula	Fingerlike projection hanging from the soft palate at the back of the mouth

V

Vacuole	Space within the cytoplasm of a cell that contains material taken in by the cell
Vas deferens	Long duct running from the epididymis to the urethra; transports sperm via peristaltic contractions
Vascular tissue	Blood; a type of connective tissue whose matrix is made of a fluid called blood plasma
Vasoconstriction	Constriction of blood vessels
Vasodilation	Dilation of blood vessels
Vasopressin	*See* antidiuretic hormone
Veins	Vessels that usually carry blood away from the tissues toward the heart
Vellus hair	Soft and downy hair that is found all over the body except the palms of the hands, soles of the feet, eyelids, lips, and nipples

Ventral   At the front of the body, in front of

Ventricle (brain)   Fluid-filled cavity in the brain; there are four of them

Ventricle (heart)   Delivery chamber of the heart that pumps blood into the blood vessels

Venule   Small vein; it drains blood away from capillaries

Vesicular ovarian follicle   *See* Graafian follicle

Vesicular transport   Type of transport across the plasma membrane of a cell in which vesicles (small sacs) carry particles

Vestibule (ear)   Central portion of the bony labyrinth of the ear; contains receptors for equilibrium

Vestibule (female genitalia)   Entire region between the labia minora, consisting of the vaginal orifice and external urethral orifice

Villi   Fingerlike projections of intestinal mucosa cells that increase the total surface area of the small intestine

Virilization   Development of masculine physical characteristics

Viscera   Organs of the abdominal body cavity

Visceral   Relating to the internal organs of the body

Vitreous body   Jelly-like substance that helps produce intraocular pressure in the eye

Voluntary nervous system   *See* somatic nervous system

W

White blood cell   See leucocyte

Z

Zygote   A fertilized ovum

Index

abdomen
    primary arteries of, 226
    primary veins of, 227
abdominal wall muscles, 114
acquired immunity, 242
active transport, 31
adenosine triphosphate (ATP), 202
adrenal gland, 270
aging
    hormones, 188
    reproductive system, 290–291
    skin, 56
agonist, 97
air sacs. *See* alveoli
alimentary canal. *See* gastrointestinal
    tract
allergen, 244
alveoli, 190, 198
    gas exchange, 199, 201
anagen, 53
anatomical
    position, 9, 10
    regions, 12, 13
androgens, 278
animal cell structure, 26–28
antagonist, 97
antigen, 242
appendicular skeleton, 74
    foot arches, 78–79
    foot bones, 78
    lower limb bones, 77
    pelvic girdle, lower limb, and foot, 77
    pelvic girdle, thigh, and leg bones, 76
    right foot bones, 78
    upper limb and shoulder girdle
        bones, 74
    wrist and hand bones, 75
arches of foot, 78–79
arm
    bone, 74
    muscles of, 121, 123

arteries, 215
    of abdomen, 226
    bronchial, 189
    of head, face, and neck, 200
    hepatic, 257
    of pelvis and lower limbs, 230
    pulmonary, 189
    of systemic circulation, 218
    of thorax, 224
    of upper limbs, 222
    and veins, 215
articular cartilage, 62
atlas (C1), 69
ATP. *See* adenosine triphosphate
atrioventricular valves (AV), 208
AV. *See* atrioventricular valves
axial skeleton, 66

back muscles, 117
bladder. *See* urinary bladder
blood, 203
    cells, 207
    clotting. *See* coagulation
    components of, 203, 204–205
    coronary vessels, 210
    flow through heart, 212, 213
    flow through kidneys, 271
    flow to heart tissue, 211
    formed elements, 205–207
    lymph and, 233
    pressure, 217
    supply in liver, 257
    supply to heart, 210
blood vessels, 215, 216. *See also*
        arteries; veins
    coronary, 210
    differences between arteries and
        veins, 215
    of heart, 217
    nephron and associated, 272
    venous valves, 216

body
    cellular organization of, 26–31
    chemical organization of, 18–25
    muscles, 99, 112, 113
    organization of, 17, 34, 43
body cavities, 14, 15
bones, 57, 88
    endoskeleton, 57
    foot, 78
    of hand, 75
    hyoid, 194
    lower limb, 77
    neck and spine, 68
    osteology, 57
    pelvic girdle, thigh, and leg, 76
    relationship of skeletal muscles
        to, 97
    remodeling, 88
    right foot, 78
    skull, 66, 67
    structure of long bone, 62, 63
    thorax, 72–73
    trunk and pectoral and pelvic
        girdles, 73
    types of, 59, 61
    upper limb and shoulder
        girdle, 74
    wrist and hand, 75
bone tissue, 57, 88
bony thorax, 72
Bowman's glands. See olfactory
        glands
brain, 148, 158
breasts, 284. See also mammary glands
bronchi, 195, 196
bronchial
    arteries, 189
    tree, 196
buccal cavity. See mouth

calcitriol, 269
cancellous bone tissue. See spongy
        bone tissue
cancer, 46
carbohydrate digestion, 266, 268
cardiac cycle, 214

cardiovascular system, 45, 203
    blood, 203–207
    blood pressure, 217
    blood vessels, 215–217, 218–231
    heart, 207–214
    hemostasis, 232
    and lymphatic system, 234
cartilage
    articular, 62
    cricoid, 194
    epiglottis, 194
    hyaline, 59, 82
    thyroid, 194
cartilaginous joint, 81
catagen, 53
cell division, 32
cellular organization of body, 26
    animal cell structure, 26–28
    plasma membrane structure, 29
    transport across plasma membrane,
        29–31
central nervous systems, 141
cerebral cortex, 156
Cerebrospinal fluid (CSF), 157
cervical vertebra (C5), 69
chemical organization of body,
        18–19
    elements of body, 21
    major compounds of body,
        22–25
chemical transmission across
        synapse, 147
chondroblasts, 59
circulation
    arteries of systemic, 218
    coronary, 208, 210
    hepatic portal, 228, 229
    primary blood vessels of
        systemic, 218
    pulmonary, 208, 212
    systemic, 208, 212
    veins of systemic, 219
coagulation, 232
coccyx, 71
collagen, 57
colon wall, 265

compact bone tissue, 58
connective tissue
    blood, 203
    bone tissue, 57
    classification, 37–39
    meninges, 156
    mesenchyme, 57, 88
    scar tissue, 138
    skeletal muscle, 92
    suspensory (Cooper's)
        ligaments, 284
    wrappings of skeletal muscle, 91
contraction, 89, 95
Cooper's ligaments, 284. See also
        mammary glands
coronary blood vessels, 210
coronary circulation, 208, 210
cranial nerves, 158–159
cricoid cartilage, 194
CSF. See Cerebrospinal fluid
cutaneous glands, 55

defecation, 264
defense mechanisms, 241
dense bone tissue. See compact
        bone tissue
depolarization, 147
dermatology, 47
dermis, 47
diaphysis, 62
diastolic pressure, 217
diffusion, 29
digestive system, 45, 245. See also
        gastrointestinal tract
disease resistance, 241
DNA molecule, 32
duodenum, 253, 263

ear, 167
electrolyte balance, 274
elements of blood, 205–207.
        See also blood
embryo, 88
endocrine glands, 173
    and hormones, 174, 175–182
    of male and female, 174

endocrine system, 44, 173
    glands, 173
    glands and hormones, 174,
        175–182
    nervous system and, 173
    and pituitary gland, 183
    stress response, 185–187
endoskeleton, 57
enteric nervous system, 268
ependymal cells, 157
epidermis, 47
epiglottis, 194
epiphyseal plate, 62
epiphysis, 62
epithelial tissue classification,
        35–36
erector spinae, 119–120
erythropoietin, 269
esophagus, 253
exoskeleton, 57
external respiration, 199
extracellular fluid, 26
eye, 164
face
    arteries of, 200
    muscles of, 104, 106–107
    veins of, 221
facilitated diffusion, 30
feces, 264
female reproductive system, 281,
        282–284
    female urinary and genital systems,
        281
    hormones, 285
    mammary glands, 284–285
    ovarian cycle, 288
    ovaries, 281
    pregnancy, 281
    reproductive cycle, 286–289
    uterine cycle, 288
    uterus with fetus, 291
fertilization, 277
fibrous joint, 81
fight-flight-freeze response. See fight-or-
        flight response
fight-or-flight response, 185, 187

fluid and electrolyte balance, 274
foot
    arches of, 78–79
    bones of, 78
    lateral view of, 77
    muscles of, 134–136
forearm
    muscles of anterior compartment of,
        123–124
    muscles of arm and, 121, 123
    muscles of posterior compartment of,
        124–125
functional bowel disorders, 268

gall bladder, 258
gametes, 277
gastrointestinal tract (GI tract), 245.
        See also digestive system
    accessory organs, 250
    bolus movement, 253
    esophagus, 253
    gall bladder, 258
    large intestine, 264–266
    liver, 256–257
    mouth, 250–252
    pancreas, 263
    small intestine, 258–262
    stomach, 253–255
    structure of, 248, 249
    in thoracic and abdominal
        cavities, 245
    walls of, 248
GI tract. *See* gastrointestinal tract
glands
    alveolar, 284
    cutaneous, 55
    endocrine, 173, 174
    mammary, 284
    olfactory, 171
    pituitary gland, 183, 184
    salivary, 250
    sebaceous, 51, 52
    sweat, 56
gluconeogenesis, 269
gluteal region, muscles of, 130
gray matter, 161

hair, 52
hand
    bones of, 75
    muscles of, 126–127
haploid cells, 277
haustra, 264
Haversian canal, 58
head
    arteries of, 200
    muscles of, 105
    veins of, 221
heart, 207, 208
    blood flow through, 212, 213
    blood flow to tissue of, 211
    blood supply to, 210
    blood vessels of, 217
    cardiac cycle, 214
    chambers of, 209
    contour of heart and valves, 210
    coronary blood vessels, 210
    coronary circulation, 208, 210
    cross-section, 209
    heart rate, 214
    intrinsic conduction system, 214
    pericardium, 207
    physiology of, 214
    pulmonary circulation, 208, 212
    rate, 214
    systemic circulation, 208, 212
hemostasis, 232
hepatic portal circulation, 228, 229
hepatocytes, 257
hip and thigh, muscles of, 128
homeostasis, 43
hormones, 188
    controlled by pituitary gland, 183
    female reproductive, 285
    fluid and electrolyte balance, 274
    not controlled by pituitary gland, 183
HPA axis. *See* hypothalamic-pituitary-
        adrenal axis
hydroxyapatite, 57
hyoid bone, 194
hypodermis. *See* subcutaneous layer
hypothalamic-pituitary-adrenal axis
        (HPA axis), 185

immune response. *See* immunity
immune system, 233, 235, 244. *See also*
    immunity
immunity, 242
immunocompetent cells, 242
immunological memory. *See* acquired
    immunity
inflammation, 242
infrahyoid muscles, 110
inspired and expired air, 199
integumentary system, 44
internal respiration, 199
intracellular fluid, 26
intrinsic conduction system, 214
islets of langerhans. *See* pancreatic islets

joints, 80

kidneys, 269, 270

lactation, 285
lacunae, 58
lamellae, 58
large intestine, 264
larynx, 194, 195
lateral vertebral muscles, 110–111
left lower quadrant (LLQ), 11
left upper quadrant (LUQ), 11
leg
    bones of, 76
    muscles of, 132–133
ligaments
    falciform, 247, 257
    suspensory, 284
lipid digestion, 267, 268
liver, 256
LLQ. *See* left lower quadrant
lobes of cerebrum, 156
lobules, 257
lower limb, 101
    arteries of, 230
    bones of, 77
    lateral view of, 77
    muscles of, 128, 129
    veins of, 231
lumbar vertebra (C5), 70

lunge, 202
lungs, 189, 196, 197
    portion of lobule of, 198
LUQ. *See* left upper quadrant
lymphatic system, 233, 235, 244
    areas drained by lymphatic ducts, 238
    blood and lymph, 233
    and cardiovascular systems, 234
    components of, 236–237
    fluid flow through body, 234
    and immune system, 45
    lymphatic vessel, 237
    lymph node, 238, 239, 240
    organization of, 235
    organs and nodules, 240

male reproductive system, 278–280
mammary glands, 284. *See* breast
    lactation, 285
    milk-secreting alveolar glands, 284
    oxytocin, 285
    prolactin, 285
    suspensory ligaments, 284
marrow cavity, 62
mastication muscles, 105, 107–108
medullary cavity, 62
meiosis, 33, 277
membrane classification, 41–42
meninges, 156
menopause, 188
menstrual cycle, 289
mesenchyme, 57, 88
metabolism, muscle, 98
metaphysic, 62
microvilli, 260
micturition, 275
micturition reflex, 275
milk-secreting alveolar glands, 284.
    See also mammary glands
mitosis, 33
motor neuron, 145, 146
    axons, 161
motor output, 139
mouth (oral or buccal cavity), 170, 250
    digestion in, 251
    papillae, 170

salivary glands, 250
taste bud, 170
taste pore, 170
teeth of adult and child, 252
tooth cross-section, 252
muscle(s), 44, 89. *See also* connective tissue
    abdominal wall, 114–115
    arm, 121, 123
    attaching upper limb to trunk,
        115–116
    back, 99, 117
    body, 99
    contraction, 89, 94–95, 96
    erector spinae, 119–120
    face and scalp, 104, 106–107
    fiber types, 98–99
    foot, 134–136
    forearm, 123–125
    gluteal region, 130
    hand, 126–127
    head and neck, 105
    hip and thigh, 128
    infrahyoid, 110
    at joints, 97
    leg, 132–133
    lower limb, 101, 128, 129
    mastication, 105, 107–108
    metabolism, 98
    neck, 108, 109
    posterior triangle, 111
    prevertebral and lateral vertebral,
        110–111
    rigor mortis, 138
    sarcomere structure, 93
    satellite cells, 138
    shoulder, 112
    shoulder joint, 116
    skeletal, 91, 92–93, 97, 102, 103, 137
    sliding-filament mechanism, 94
    spinocostal, 119
    spinotransversales group, 120
    suprahyoid, 109–110
    thigh, 130–131
    thorax, 114
    tissue, 40, 89, 90
    transversospinales group, 120
    trunk, 112, 117, 118

    upper body, 112, 113
    upper limb, 100, 121, 122
myelinated fiber, 145
myelin sheath, 145

nails, 54
nasal cavity, 192
neck
    arteries of, 200
    bone, 68
    muscles of, 105, 108, 109
    veins of, 221
nephron, 270, 272
nerve(s), 148
    brain, 158
    cells, 34
    cervical, 12
    classification of, 148
    cranial, 158–159
    impulse, 94, 147, 161, 164, 185
    spinal, 162–163
    structure of, 148
    vestibulocochlear, 167
nervous system, 44, 139
    brain, 148–157, 158
    cells, 143
    central and peripheral, 141
    chemical transmission across
        synapse, 147
    comparison of endocrine and, 173
    cranial nerves, 158–159
    divisions of, 140
    ear, 167–169
    effects of sympathetic and
        parasympathetic, 142–143
    and endocrine systems, 173
    enteric, 268
    eye, 164–166
    functions of, 139
    motor neuron structure, 145, 146
    motor output, 139
    mouth, 170
    myelin sheath, 145
    myelinated fiber, 145
    nerve impulse transmission, 147
    nervous tissue, 143–144
    neuroglia, 143, 144

neurons, 143, 144, 172
nose, 171, 172
organization of, 140
reflex arc, 140
sense organs, 164–172
sensory input, 139
sensory receptors, 139
spinal cord, 160–163
nervous tissue, 143
    neuroglia, 143, 144
    neurons, 143, 144, 172
    neurons classification, 144
nose, 171, 172
    skeleton of, 192

olfactory glands, 171
olfactory receptors, 171
oral cavity. See mouth
organization of body, 17
    cellular, 26–31
    chemical, 18–25
    levels of structural, 17
    life cycle of cell, 32–33
    system level of, 43–45
    tissue level of, 34–42
organs, 43
osmosis, 30
osseous tissue. See bone tissue
osteology, 57
osteon canal, 58
ostia, 193
ovarian cycle, 288
ovaries, 281
oxytocin, 285

pancreas, 263
pancreatic islets, 263
papillae, 170
paranasal sinuses, 193
parasympathetic nervous systems,
    142–143
passive transport processes, 29
pectoral bone, 73
pelvic girdle, 73, 76
pelvis
    arteries of, 230
    veins of, 231

pericardium, 207
periodic table, 20
periosteum, 62
peripheral nervous systems, 141
peristalsis, 253
peritoneal cavity, 247
peritoneum, 247
pharynx, 192, 193
pH scale, 20
pili. See hairs
pituitary gland, 183, 184
planes, 16
plasma membrane, 29–31
platelet plug, 232
plexuses, 163
pregnancy, 188, 281
prevertebral and lateral vertebral
    muscles, 110–111
prime mover, 97
prolactin, 285
protein digestion, 267, 268
puberty, 188
pulmonary
    arteries, 189
    circulation, 208
    respiration. See external respiration
    and systemic circulation, 212
    ventilation, 198–199

quadrants, 11

reflex
    arc, 140
    micturition, 275
refraction, 164
reproductive system, 45, 277, 292.
    See female reproductive system;
    male reproductive system
    aging and, 290–291
    female reproductive system, 281–284
    fertilization, 277
    gametes, 277
    haploid cells, 277
    male reproductive system, 278–280
    meiosis, 277
    uterus with fetus, 291
    zygote formation, 277

resistance reaction, 185, 186
respiration, 199
respiratory system, 44, 189
    alveoli, 198
    bronchi, 195, 196
    bronchial arteries, 189
    bronchial tree, 196
    changes in thoracic cavity during
        breathing, 200
    composition of inspired and expired
        air, 199
    external respiration, 199
    functions of, 189
    gas exchange, 201
    hyoid bone, 194
    internal respiration, 199
    larynx, 194, 195
    lungs, 189, 196, 197
    nasal cavity and pharynx, 192
    nose, 192
    ostia, 193
    paranasal sinuses, 193
    pharynx, 193
    physiology of, 198
    portion of lobule of lungs, 198
    pulmonary arteries, 189
    pulmonary ventilation, 198–199
    trachea, 195
    zones and structures of, 191–198
right lower quadrant (RLQ), 11
right upper quadrant (RUQ), 11
rigor mortis, 138
RLQ. See right lower quadrant
RUQ. See right upper quadrant

sacrum, 71
salivary glands, 250
sarcomere structure, 93
satellite cells, 138
scar tissue, 138
sebaceous glands, 51, 52
sense organs, 164
    ear, 167–169
    eye, 164–166
    mouth, 170
    nose, 171, 172

sensory
    axons, 161
    input, 139
    receptors, 139
serotonin, 232
shoulder
    girdle bones, 74
    joint muscles, 116
    muscles, 112
sinusoids, 257
skeletal muscles, 91, 102, 103, 137
skeletal system, 44, 57
    appendicular skeleton, 74–79
    axial skeleton, 66–73
    bones, 57–59, 88
    joints, 80–87
    organization of skeleton, 64–65
    types of bones, 59, 61–63
skin, 47, 48
skull bone, 66, 67
sliding-filament mechanism, 94
small intestine, 258–262
smell and taste, 171
sound waves, 167
spinal cord, 160
spinal meninges, 161
spine
    bone, 68
    curves of, 71
    spinocostal muscles, 119
    spinotransversales group, 120
spongy bone tissue, 59
stomach, 253
    digestion in, 254–255
    and duodenum, 253
    wall, 255
stress response, 185
subcutaneous layer, 47
subcutis. See subcutaneous layer
superficial fascia. See subcutaneous layer
suprahyoid muscles, 109–110
suspensory (Cooper's) ligaments, 284
sweat glands, 56
sympathetic nervous systems, 142–143
synapse chemical transmission, 147
synergists, 97

synovial joints, 82
systemic circulation, 208
   abdomen arteries, 226
   abdomen veins, 227
   arteries of, 218
   blood vessels of body, 200–227, 230
   head, face, and neck arteries, 200
   head, face, and neck veins, 221
   hepatic portal circulation, 228, 229
   hepatic portal venous circulation, 228
   lower limb and pelvis arteries, 230
   pelvis and lower limb veins, 231
   thorax arteries, 224
   thorax veins, 225
   upper limb arteries, 222
   upper limb veins, 223
   veins of, 219
system level of organization of body,
   43–45
systolic pressure, 217

taeniae coli, 264
taste, 170
teeth, 252
telogen, 53
tensile strength, 57
testes, 278
thigh
   bones, 76
   and hip muscles, 128
   muscles, 130, 131
thoracic cavity during breathing, 200
thoracic vertebra (T6), 70
thorax
   arteries of, 224
   bone, 72–73
   muscles of, 114
   veins of, 225
throat. See pharynx
tissue level of organization of body, 34
   connective tissue, 37–39
   epithelial tissue, 35–36
   membranes, 41–42
   muscle tissue, 40, 89
tissue respiration. See internal respiration
tissue types, 34–35

tonus, 138
tooth, 252
trabeculae, 59
trachea, 195
transversospinales group, 120
trigone, 275
trunk
   bone, 73
   muscles of, 112, 115–118

upper body muscles, 112, 113
upper limb, 100
   arteries of, 222
   bones, 74
   muscles of upper limb, 115–116,
      121, 122
   veins of, 223
ureters, 274, 275
urethra, 275
urinary and genital systems, male, 278
urinary bladder, 274, 275
urinary system, 45, 269
   female, 281
   kidneys, 269, 270–271
   male, 278
   nephron, 270, 272–274
   organs of, 269
   ureters, 274, 275
   urethra, 275
   urinary bladder, 274–275
urine analysis, 271
urine flow, 271
uterine cycle, 288
uterus with fetus, 291

veins, 215
   of abdomen, 227
   arteries and, 215
   bronchial, 189
   of head, face, and neck, 221
   hepatic portal, 257
   of pelvis and lower limbs, 231
   pulmonary, 189
   of systemic circulation, 219
   of thorax, 225
   of upper limbs, 223

venous valves, 216
ventricles, 157
vertebral
    column, 68
    muscles, 110–111
vesicular transport, 31
vestibulocochlear nerve, 167
villi, 260
visceral
    effectors, 185
    peritoneum, 247

voice box. *See* larynx
Volkmann's canals, 58

water, 276
white matter, 161
white thrombus. *See* platelet plug
windpipe. *See* trachea
wrist, bones of, 75

zygote formation, 277